Understanding Evolution and Creation!

Patrick C. Marks

Markarian Publishing
Phoenix, AZ

Understanding Evolution and Creation!

Cover art by Amy Weller: wellerart@gmail.com

Interior illustrations by Ron Zeilinger.

ISBN: 978-0-9832097-4-4

For information on booking speaking engagements with the author, please contact: **pat@fourteensix.com**

Printed in the United States of America

Acknowledgements

As a public school educator it has been my task to take complicated scientific concepts and repackage them in a manner the average 8th grader can understand. *Understanding Evolution and Creation* is a continuation of this task. This book is a summary of Evolutionary concepts and the scientific reasons Evolution theory falls short. I make no claim to be a professional scientist but lean instead upon the writings and authority of credentialed Creationist scientists.

Special Thanks

I would like to thank my editors Bonnie Halley and Jody MacArthur for their outstanding efforts. Special thanks to my friend Stephen Opgenorth for formatting and layout help. Thank you Amy Weller and Ron Zeilinger for illustrations. Thank you James Marks for many illustration ideas – your sense of humor is "*special*". I am indebted to the late Dr. Duane T. Gish but not just for the wonderful foreword for this book – your ministry directly contributed to bringing me to Christ. Thank you to Bill Wilson for giving your endorsement but also for encouraging us in ministry. I am especially grateful to Mark Hoffman for the introduction but also to his brother Dave and the rest of the Foothills Christian Church family – your support is a blessing I cannot adequately describe with words. Thank you to Bill McKeever of mrm.org – you are a true friend and fellow-soldier for the King.

Thank you to Dennis Muller – without you the original book and this revision would never have been inspired in the first place.

Finally, but most importantly, this book is dedicated to Melissa Marks: *Your worth to me is far above rubies. The heart of your husband safely trusts in you...* (Proverbs 31:10,11)

Foreword

"...I think a case can be made that faith is one of the world's great evils, comparable to the smallpox virus but harder to eradicate."

Richard Dawkins (1996 Humanist of the year).
"Is science a religion"? The Humanist 57:26 (Jan/Feb 1997). Print.

Richard Dawkins is a famous biologist, evolutionist and atheist. He is, of course, one of those who ridicule Creationists. It has been reported that almost 70 percent of our young people do not return to their church after attending a secular university. If, however, young people and others read this book by Patrick Marks, the result can be dramatically different. This book exposes the weaknesses and fallacies of Evolution theory and at the same time provides a wealth of scientific evidence that supports Creation.

In this book, Marks provides a thorough treatment of practically every important subject related to origins, from the origin of the universe, the origin of life, the origin of man, the fossil record, thermodynamics, dinosaurs and many others.

In contrast to most evolutionists who treat their opponents rather unkindly, Patrick Marks approaches each subject with care to listen and consider the opponents position.

I highly suggest to all of those who are interested in the Creation/Evolution controversy to obtain a copy of this book. It is one of the best I have seen. You will enjoy and profit from every chapter.

Duane T. Gish, Ph.D

Senior Vice President Emeritus, Institute for Creation Research,
Dallas, TX
September 30, 2010

Table of Contents

Question 1

Question 2

Question 3

Question 4

Question 5

Question 6

Question *14*

Question *15*

Question *16*

Question *17*

Question *18*

Question *19*

Introduction

Perhaps the most destructive idea that can enter a young person's mind is the idea that there is no reason or purpose for their life. It is dangerous to believe life is a simple accident of circumstance and that there is no destiny or purpose to live for and discover. Along with this destructive belief, young people are also told there is no secret or pattern to the world they find themselves in, no way of knowing what is the right way to play the game and no instruction manual or rule book for life. In short, they are led to believe there is nothing better to do than become the slave of whatever pleasure each moment brings.

This, however, is exactly what young people are told over and over every day in our public schools and universities. They are taught that they are products of purposeless Evolution in a meaningless universe and no other views are allowed to be considered. Why? Why are these educational institutions so afraid of a view other than Evolution?

The result of this indoctrination surrounds us for all to see. Young people are confused, aimless, apathetic, lacking in self-restraint and often self-destructive. And all this results from believing something that is not true.

As the founder and executive director of several youth organizations and two schools, I have met thousands of youth over the past 30 years. I have watched the increasing demoralization of our young people over these years as the ideas of Evolution have increasingly taken hold in our culture.

But, I have also watched life and hope flood into the eyes of youth when they learn that they are not accidents but are instead special creations of a wise and loving Being. This is why our mentoring program to *at risk* youth begins with establishing the concepts of design and purpose that flow out of the truth of special design by a Creator. We begin with this all-important understanding because I have not found any other truth that so effectively begins to change a young person's perspective on life. First understanding they are a special creation by God builds hope in them.

Patrick Marks has given us a wonderful gift in this book *"Understanding Evolution and Creation"*. I have not seen a better presentation of the case for Special Creation aimed at youth than what you will find in this book. Believe me; young people are hungry for this vital truth and in desperate need of the information in this book. I witness this need almost every day.

Share this book with a friend or better yet, master the information in it so you may be able to share the truth effectively with your classmates. My prayer is that this book will find wide circulation in Christian schools, churches and youth organizations. I know that we will use it widely. But most of all, I pray that this book will find its way into the hands of young people who are in need of its answers.

Mark Hoffman

Founder: Youth Venture Teen Centers.

Co-Senior Pastor, Foothills Christian Church, El Cajon, California.

Chapter 1

Why should you read this book?

It isn't easy being you, is it? Lots of questions! Some people just give up and quit caring about what they do in life every day. You know the type: they don't seem to care about anything – so why *should* you care? It seems a lot easier to just have whatever fun you can and not worry about the meaning of life. But everybody has to figure out how to get by in life eventually.

The truth is that what you believe about where you came from has a lot to say about who you are and what should matter in your life. It can change how you see everything about your life. Imagine if you were an orphan and you had always been told that your parents had been unimportant people who died in a car crash. What if you found out your parents had actually been heroes who died saving many lives? It could really change your attitude about everything.

But it can be pretty confusing when every science class you take and every popular magazine like National Geographic tells you that your ancestors were a group of apes sitting on the savannah picking fleas off each other a million years ago. And if you're really just an animal, then there's no sense worrying about what you do with your life. After all, chimps don't worry about playing fair. They pretty much do whatever they feel like doing, so why shouldn't we do just the same?

Many people think they are just accidents or no different than other creatures. If your parents were too busy messing up their own lives to care very much about yours, it can make you feel like you're just an accident too and life can seem pretty pointless. Even if people around you really care about you or you're involved with someone who really loves you, you can still wonder if your life has any real meaning. Why not just have a good time and do whatever you want? After all, that's what the chimps do!

Many scientists and teachers today believe that life got started about 1.6 billion years ago in a pool of slime. Just by accident, simple life forms became more and more complex over millions of years until life forms started walking and talking and teaching 10th grade biology classes. This belief is called the "*Theory of Evolution.*"

Other scientists believe that life is too complex to just come about by accident. They believe life got started by an intelligent designer. Many people who believe in Intelligent Design think that the great designer is the God of the Bible. This idea is called "*Scientific Creationism*".

Evolution is what public schools teach the kids and many scientists, teachers and most textbooks teach Evolution as if there is no question about it. Many teachers will say there is some debate about how Evolution happened or how fast or what path it took but they don't question the actual idea of Evolution. In fact, many universities and school districts will threaten or even fire people who say Evolution might not be true. One very famous scientist, Dr. Richard Goldschmidt said, "*Evolution of the animal and plant world is considered by all those entitled to judgment to be a fact for which no further proof is needed.*"[1]

If Evolution is really how we came to be alive today, then people are just very smart animals and can make up their own rules about how to live just like animals do. In fact, animals live by the rule, "*Whoever is the strongest or fastest gets to live and everyone else becomes lunch,*" so there really is no such thing as a real right and a real wrong. Right and wrong is just whatever we make up in our own minds.

On the other hand, if Intelligent Design is true, then being alive today is the result of someone else's plan. Creationists believe this designer is the God of the Bible and because God made everything, He has the right to make the rules about what is right and what is wrong. Either way, both ideas may have an effect on how you decide to live your life. This is why it's important to decide what *you* think.

This book will show you how the real facts of science line up better with Creationism than any other idea. But don't worry about any complicated scientific words or super technical ideas. This book is designed to be easy to read, easy to understand and easy to remember so that when you hear about Evolution, you will already know what the Bible says and why the Bible makes more sense.

To make this simple, we will begin by following the phrase, "*Evolution isn't science, it's science FICTION.*" Every letter of the word "*fiction*" lines up with an important topic about Evolution and Creation. There are also 23 questions and answers about all sorts of science topics so you can skip around to what interests you. Each section is short enough to read and understand in just a few minutes and after each section, you will find a phrase or a word to help you remember the information.

The real facts of science say you are not an accident and your life matters very much. By the time you finish reading this book, you should be able to see why most of the world's greatest scientists in the past and a great many scientists today think life had to have been made by God. After that, it's up to you. How are you going to let this information change how you view your life?

Chapter 2

Evolution: What are many scientists thinking?

Pretty much every science book you are likely to see in a public school or secular college classroom assumes the idea of Evolution is a fact of history. Most of these books tell you that the best and the brightest scientists today all believe in the Evolution story. They may even tell you how once upon a time people believed in the Bible's story about how the world and people came to be, but now we know better.

In order to understand Evolution, we have to be clear what we are really talking about in the first place. The word evolution itself just means "*change*". No one really questions the idea that plants and animals change over time. For example, the average height of human beings has gotten taller over the last few hundred years. But there is a big difference between this kind of change and the idea that ape-like creatures living in trees changed slowly into human beings. This idea makes it look like someone's trying to make a monkey out of you!

So, how did plants and animals get here? This is the big question and modern Evolution is the idea that living things got here by a natural process. It is supposed to have started about 1.6 billion years ago when the first simple form of life got started, by accident, in a chemical soup. This first form of life was just a molecule. After millions of years, these first molecules changed into cell life. Over many more millions of years, just by mistakes between the parent cell and the daughter cells, these cells changed into every kind of animal and plant we see around us today.

This idea isn't really new but it became popular with scientists after Charles Darwin published his book called "*On the Origin of Species*" in 1859. Darwin got his ideas for the book during a trip he took around the world on the ship *HMS Beagle*. Darwin's job on the voyage was to study and describe animals and plants. He noticed that some animals, such as a tiny bird called the Galapagos finch, had special features, also called adaptations, which allowed them to survive in different places. One kind of finch had a beak designed for catching insects but another finch had a beak built for eating seeds. He wondered why the finch beaks had changed from one kind to another.

Darwin came up with an idea called "*natural selection*" that he believed could explain why the finches had different kinds of beaks. Natural selection means that some creatures survive because they have special features allowing them to compete for food or mates better than others. The more "*fit*" creatures are able to survive and pass on their good traits to the next generation. Darwin thought this process could cause one kind of creature to completely change into a different kind of creature over great lengths of time.

Evolution is supposed to be a very slow process, something that won't be noticed even over a lifetime. No one has ever seen one kind of creature completely change into a different creature. Evolutionist scientists can only really say Evolution *might* be true because real science is supposed to be about natural processes we can watch, measure and predict. One very famous Evolutionist, George Gaylord Simpson, said, "*Statements that cannot be checked by observation are not really about anything...or at the very least, they are not science.*"[2] Clearly, we can't decide if Evolution is what really happened by watching it.

The only way to decide would be to look at all the evidence from the past, such as the fossils of animals and plants, and see if this evidence fits with the Evolution idea. If the evidence doesn't fit, we will have to change the idea or come up with another idea that fits the facts better. This is what science is really about – checking an idea (also called a *hypothesis*) against the evidence, making predictions and seeing if the facts fit or not. The truth is ideas like Evolution are actually "*scientific models*".

If the Evolution model is what really happened in the past, we should see three things in nature: (1) There should be thousands of examples of interesting creatures partway between one kind of creature and another. Partway forms are called *transitional forms.* (2) The Earth should be very, *very old* because Evolution is supposed to be such a slow process. (3) We should be able to see how the very complicated systems we see in creatures today came into place by some *step-by-step process.*

If Evolution is what really happened in history, then all three of these points must be obvious in the natural world. If we don't see these three things, it is strong evidence that Darwin's idea isn't what really happened.

It isn't just scientists and teachers who should care whether or not Darwin's idea is true. If Evolution is what really happened, then all of us are just animals. And animals don't care very much about right and wrong! Darwin himself said that since people are just animals, they should live their lives based on the strongest impulse or instinct that they feel.[3] Think about that for a moment. If a person's anger is the strongest impulse, they might murder someone – but if we are all just monkeys anyway, what would it matter?

How to remember this:

We have not arose, from Monkey TOES.

T.OE.S.

- **T** = TRANSITIONAL FORMS – there should be many examples of life-forms in-between one form and another.
- **OE** = OLD EARTH – this process should take many millions of years.
- **S** = STEP by step – every life-form alive today should be explained by a natural step by step process.

Chapter 3

Creation: What is the opposite idea to Evolution?

Evolutionary science tries to show how life-forms could have come from nothing by some natural process. Creation science tries to show that everything could not have come about by any natural process.

Many scientists think that including God in the creation story is not science. The problem with this thinking is that science is supposed to be about finding out how things work or change, not deciding in advance that God cannot be part of the answer. We study nature "*scientifically*" by asking questions, coming up with possible answers and doing experiments to see if our possible answers are right or not. This is called the "*scientific method*". The word "*science*" itself means "*knowledge*". Real science should be about following the evidence wherever it leads - even if it leads to something, or *someone*, we didn't expect.

When we do follow the evidence we find many examples of life-forms that could not have been built by any natural process. We find that most life-forms are incredibly complex. Most of them have so many inter-locking parts only an intelligent agent could have put it all together. The truth is, there are just too many examples of design in nature for Evolutionists to ignore.

The Bible says, "*In the beginning, God created the heavens and the earth...*"(Genesis 1:1). Many very qualified scientists, such as Dr. Henry Morris and Dr. Duane T. Gish, think the evidence clearly supports God as Creator. Many Creation scientists once believed in Evolution but they were willing to follow the evidence wherever it led. After years of studying fossils, living animals, plants, the solar system, stars and other systems, they became confident only God could have made it all.

The theory of Creation science teaches that God created the basic kinds of plants and animals, in the very beginning, using only His own power. God is not creating anything today because He designed living things to survive in a wide range of places. Creationists agree that animals and plants will change over time but Creationists believe these changes will never result in one kind of animal changing into another. Dogs and cats are clearly different kinds of creatures. We do not see dogs slowly changing into cats – no "*dats*" – or cats changing into dogs – no "*cogs*".

Just think about dogs for a moment. There are many different kinds of dogs, like poodles and German shepherds, but they are all still dogs. Some dogs with long, thick hair are able to survive in colder areas. Other dogs with short hair are better able to live in places with warmer climates. Dogs come in all sorts of varieties and sizes too but they are all clearly the "*dog-kind*". Creationists call this wide range of differences in animals such as dogs "*variation within a kind*". They believe that God created this ability to adapt so they could survive in many different areas and climates on Earth. This ability to change would be especially important for animals if the weather became a great deal warmer or colder.

No one has ever seen one animal-kind change into another. After thousands of years of farming, where farmers have worked to get the best milking cows or the largest sheep, cows are still cows and sheep are still sheep.

On the other hand, Creationists believe the creation of nature was a one-time event. This means Creationists and Evolutionists have a similar problem - no human scientist was around to witness what actually happened. In fact, one Evolutionist said that belief in both theories is really the same sort of thing. Each one has many supporters who are certain they are right, but neither group can come up with any absolute proof.[4]

We can only decide what happened based on what we see happening around us today and the evidence from fossils. In the last section, you learned there are three basic predictions scientists should be able to see in nature if Evolution is what really happened in the past. These are in-between forms, also called transitional forms; proof that the Earth is billions of years old; and a believable step-by-step natural process that would show how creatures were built over time.

WOW! Did you see him make that Platypus?! THAT'S going to give the Evolutionists a real head-ache Someday!

Creation science also has some basic predictions about nature scientists should be able to find if Creation is true. First, all the major kinds of plants and animals should be found fully-formed right from the very beginning. There should be *no in-between, or transitional forms*, between the major kinds of plants and animals. This doesn't mean that God created every type of dog, like shepherds and retrievers, directly. Instead, God created the *dog-kind* with an ability to branch out into a wide variety of dogs. If Creation science is true we should *not* see the dog-kind slowly turning into another kind such as the horse-kind over the years. Second, living things should be able to change, or adapt, to a wide range of environments without changing into new kinds.

Finally, if Creation science is true there should be many examples of very complex creatures that cannot be explained by any step-by-step natural process. In other words, there will be examples of living creatures and systems that can only be explained if someone had planned and built them. All of these things should be clearly seen in the natural world if God is the Creator as the Bible teaches.

It is important to decide what you believe and even more important to know why you believe it. Evolution says you are nothing more than a very smart animal. And animals do whatever they want, whenever they want! So Evolution basically says you may as well make up your own rules in life.

Creation science teaches that people are a special Creation made by God and that God told us in the Bible how and why He made everything. In fact, the Bible says that God's power can be seen every day in what He has made.[5]

How to remember this:

God spoke and then the Earth appeared, the FANS in heaven really cheered.

F.A.NS.

- **F** = FULLY FORMED – life-form *kinds* should never change into completely different kinds.
- **A** = ADAPTABLE – life-form kinds should be able to adapt to different environments without changing into a completely different kind.
- **NS** = NO natural STEP-by-step process can account for how life-forms were built.

Chapter 4

Science: What is it, really, and why should I care?

Science isn't just a class you don't really want to take but it isn't just something that only matters to scientists either. In today's world, science can make an idea look good and that kind of power can affect your life.

This is true because most people think of something that is "*scientific*" as being true, whether we like it or not. Just think about all the current hysteria over "*global warming*." Fear of global warming is leading to many new laws restricting all kinds of freedoms. When science says these activities supposedly cause the Earth to warm up, we believe it. Pretty soon it may become illegal to mow the grass in your front yard or drive a car for fear of adding to the global warming problem. But if the "*science*" of global warming is not accurate, these new rules would not be needed.

Many people think the theory of Evolution is science and is therefore true. In fact, many scientists, teachers, professors and people in the media never question Evolution. They admit there may be some debate about how and when Evolution happened, but since many people see Evolution as scientific, they don't question it.

What we believe is important because if people are just animals, we can make up any rules about life that we want. On the other hand, if Evolution is not true, then we are specially designed creatures. And our designer has something to say about how we live our lives!

Neither Evolution nor Creation are completely scientific in the strict sense of the word. This is true because science is mostly about describing facts by observing the world around us and no one has ever observed Evolution or Creation. No one has ever seen one kind of animal or plant change by a natural process into a completely different kind of animal or plant. But no one has seen God create anything new either. The truth is, Evolutionists can be just as religious as Creationists.

Some people may say that watching a sample of bacteria develop a resistance to a drug is an example of Evolution. This really isn't Evolution because the bacteria are still bacteria. Others argue that scientists have forced fruit flies to be born without legs or with four wings by exposing the flies to radiation or by changing their DNA. But this is not Evolution either because the flies are just mutants. This isn't an example of a natural process either. In fact, it shows just the opposite because the scientist is an "*intelligent designer*" that is intentionally working on changing the fruit fly.

The dictionary says science is simply organizing knowledge. It is supposed to be a way for people to figure out an accurate description of the world and how it works. Scientists can get this knowledge from observing the natural world and from experiments. Scientists generally make up experiments to try out their ideas and then they write down what happens. Later, they can organize what they learn and come up with general rules or laws about how things work in the natural world. This is important because understanding how things work can help us invent medicines and better ways to feed people or avoid terrible plagues.

The idea behind science is to try and discover facts about the natural world using a very precise checklist called the "*scientific method*". Scientists want to use this method to be certain the facts they discover are really genuine. They want to keep their own opinions and background ideas from clouding up their observations. In this way, scientists should be able to find an accurate and reliable view of the world and how it works.

The scientific method should have at least four parts. The first step is to watch and describe a natural process. The second step is to come up with an idea to try and explain the natural process you observed. This is called "*forming a hypothesis*". The third step is to use the hypothesis to predict or guess what may happen with the process over time. Finally, scientists should do experiments to see if the predictions they made really come through.[6] Experiments that show the predictions are true need to be reviewed by other scientists in order to make sure the experiments were not contaminated or the original observations were not biased. This is how science is supposed to work.

Evolution cannot be studied like this because no one can live long enough to watch an experiment where one kind of animal gradually changes into another kind of animal. On the other hand, no one can see God create something out of nothing either. Neither theory can be tested in a laboratory.

Both Evolution and Creation are actually scientific models. A model is an idea based on the facts scientists have studied about how something <u>might</u> have happened. A model can never be completely proven, because no one can go back in time and watch Evolution or Creation take place. All we can do is study the facts and see whether the facts fit into the model. If most of the facts fit into our model, then we can feel pretty confident that our model is true.

In the end, we can see that Evolution is no more scientific than Creation science. We must decide "*which model fits the facts best?*" This is why we should not be fooled into believing Evolution is true just because it is supposedly scientific.

How to remember this:

The nature of real science – is seeing which SIDE has the facts in compliance.

S.I.D.E.

- **S** = SEE it (we observe things first)
- **I** = IDEA (we form an idea called a Hypothesis)
- **D** = DECIDE (we decide what might happen if we change something. This is called a prediction)
- **E** = EXPERIMENT (we experiment to see if our predictions are true)

Chapter 5

Theistic Evolution: Maybe God used Evolution to make the world?

Some people get worried because it seems like so many scientists, teachers and people in the media believe Evolution is the scientific truth. No one wants to feel like they believe in something stupid. This may be why some people wonder if both Evolution and Creation can be true at the same time. Maybe God guided the process of Evolution or maybe we just don't understand what the Bible really says about how the world was made. At first glance, this seems like a good idea. But there are some very big differences between these two ideas that just don't mix.

Evolution, for example, is a theory that tries to explain everything naturally without any help from God. Evolutionists do not try to fit God into their model. There are some Evolutionists who think Creationists are not really scientists at all. In fact, it is often Christians, not Evolutionists, who try to fit these two theories together.

On the other hand, the Bible's story of Creation tells us that God created everything out of nothing by His own power. The Bible does not try to fit Evolution into the Creation model. In fact, the Bible clearly says in Genesis Chapter One that God created the world in six days. It doesn't say anywhere that God used anything like Evolution to make the world.

It is very important to first understand that there are many well-meaning honest Christians who think God might have used Evolution to make the world. How we think about the Creation of the world is not what makes a person a Christian. In fact, a

Christian is a person who believes that humans are separated from a relationship with God because of the evil things we have done (our sins). Christians believe Jesus Christ died on the cross to pay the price for our sins and He rose again from the dead. This proves He is God. Any person who puts their trust in Jesus to forgive their sins and turns around from living a life that doesn't care about sin is a Christian. So we can have different ideas about the book of Genesis and still be Christians.

Still, most Christians do believe that God created the world just as it says in Genesis. In fact, the writers of the New Testament either quoted or referred to Genesis 165 times. Not one of these people in the New Testament, including Jesus Himself, ever hinted that Genesis shouldn't be understood just as it is written.[7]

When the ideas of Evolution and the story of Creation are mixed, it is called "*Theistic Evolution*". There are many different kinds of Theistic Evolution, but the two most popular are the Gap theory and the Day-Age theory.

"Irving discovers that cats and water really don't mix after all"

The Gap theory says God actually created the earth twice. The first Creation lasted for many millions of years. During that time, all the fossils and sedimentary rocks we have today were made. Later, a huge disaster destroyed the world and God had to re-create it in six days as it says in Genesis. This theory attempts to put several billion years between the first two Verses of the Bible.

The truth is, however, there is no gap between Verses One and Two in the first chapter of Genesis. This is true because these Verses were originally written in the Hebrew language. In Hebrew, Verse Two is included in Verse One. They are not separate. In fact, the Verse numbers we have today were not added into the Bible until thousands of years after they were first written down. In other words, the first Verse and the second Verse are one complete idea in Hebrew, not two ideas with a huge unwritten gap between them.

The second popular type of Theistic Evolution is called the Day-Age theory. In this theory, some writers try to prove the six days described in the first two chapters of Genesis were actually millions of years long. These people say the Bible is just lumping all the millions of years of history into six different pieces and then calling each piece "*one day*".

There are two major reasons, among others, that the Day-Age theory doesn't work very well. The first reason is that the order of events in the Genesis model does not match the order of events described by the Evolution model. They are completely different. Second, the Hebrew word for "*day*", which is used over 700 times in the Old Testament of the Bible, never means anything other than a single, 24 hour day.[8] [9]

Another reason both the Gap theory and the Day-Age theory don't work very well is the Bible makes it clear there was no death before Adam and Eve disobeyed God in the Garden of Eden. If Evolution were true, then there would be millions of years in which hundreds of millions of animals would have suffered and died. The Bible does not try to fit Evolution into the story of Genesis. Only people try to make these two ideas fit!

Some very well-meaning Christians who believe in some form of Theistic Evolution really shouldn't worry about trying to mix these two ideas. They ought to look carefully and see that the Bible doesn't support mixing the two. Besides, as you will see later in this book, the evidence showing that the Evolution model is false is very, very strong. In fact, you will see that most of the evidence from fossils and the study of living things clearly fits the Creation model - just as it is written in Genesis. As a result, there is no need to try to mix Evolution and Creation.

How to remember this:

Some think that Genesis needs a fix - but Evolution and Creation do NOT mix.

N.O.T.

- **N** = **N**O death before Adam sinned.
- **O**= **O**RDER of days does not fit order of Evolution.
- **T** = **T**OGETHER, Genesis Verse One is together with Verse Two, there is no gap.

Chapter 6

Evolution is science F.I.C.T.I.O.N: How to see the big picture on a little screen.

The scary thing about time is that it only moves in one direction. Everything you say or do is immediately lost in the past. Do you remember the last time you said something you wished you hadn't said? Even if you say you are sorry right away, it's too late. Once something is in the past, it is gone forever. Something that you did or said yesterday morning is as far away from your touch as the dinosaurs.

Since the past is just a memory, what really happened can be lost or forgotten very quickly. This is why historical records are so important. How would we know anything about George Washington if we didn't have what he said and did written down for us to study today? Even when people write down what happened during their lives, those records can be unfinished or only tell part of the story. Worse, records can be lost or broken in some way so that sometimes we can't really know for sure today what happened in the past.

In a court of law, finding out what happened in the past is especially important. Witnesses tell the court what they saw or heard. Experts talk about evidence such as fingerprints or DNA. It is up to the people of the jury to put all the pieces of the puzzle together and decide what really happened. This is what it means to try and "*see the big picture based on the small pieces*". But the jury cannot ever be 100% certain because they were not witnesses. They must add up all of the evidence and decide what is "*beyond a reasonable doubt*". In other words, they have to make a decision based on what makes sense from the evidence.

We cannot go back in time to see what actually happened when the world was made. Instead, we have to decide if the Evolution or the Creation model makes more sense based on the evidence. Just like a jury, we cannot get 100% of the picture, so it is important to think about the most important evidence. We need to look at the parts of the puzzle that really let us see the whole picture. Of course, you can see a whole image even if some minor parts are missing. In the same way, we can see the image of the past if we look at the most important pieces of the puzzle.

You can remember these big puzzle pieces for the Evolution versus Creation debate from the letters of the word "*FICTION*". The letter "*F*" reminds you of the word "*Fossils*". Fossils are the hard remains of plants and animals. If living things changed from one major kind into another over many millions of years, the fossils should show examples of thousands of strange, in-between or transitional fossil creatures. If Creation is the picture from the past, the fossils should not show any transitional forms and all of the major kinds will be seen complete right from the start.

The "*I*" in FICTION will remind you of the words "*Irreducible Complexity*". Something that is "*irreducibly complex*" is something very carefully put together. If Evolution is true, every natural system or creature we see should be explained by a lucky, step-by-step process. But if we see systems or creatures so carefully put together there is no reasonable way they could be put together by accident, then Evolution is not the picture of the past.

"C" is for "*Chance*". It is one thing to talk about how something might have happened, but what are the chances something can happen based on real math? So what are the chances that living things could just happen? If the chances are good that lucky breaks could make living things that would be good for Evolution. If the chances are very poor for Evolution, then Creation is much more likely to be the picture of the past.

The "*T*" in FICTION is about how energy is used and how things are built in nature. Every living thing requires energy to survive, so where did all the energy come from and what is going to happen to it over time? To answer this question, we have to look at "*Thermodynamics*". Thermodynamics is a part of the science of physics that describes how energy is used in the universe.

The "*I*" in FICTION is for "*isotope*". If Creation is true, the Earth can be any age, old or young - it doesn't really matter - but Evolutionists must show that the Earth is billions of years old. Most Evolutionists use "*radio-isotope dating*" to show the age of rocks. It might not matter how old the Earth is for Creation, but if the isotopes don't show the Earth to be billions of years old, then the puzzle pieces don't show a picture of Evolution.

"O" is for "*Old Earth*". There are Creationists who think the Earth is millions of years old and others who think it is much younger – but the point is there are many other puzzle pieces besides isotopes showing that the Earth cannot be as old as it needs to be for Evolution.

Finally, the "N" is for "*Noah's flood*". The Bible tells the story of a world-wide flood, as do many ancient stories from all over the Earth. If the Flood really happened, we should be able to explain it and see the results of the Flood in nature. We should also be able to show scientifically how the world today came from Noah and the animals on the ark.

There are certainly other questions people have about the past when they wonder about Evolution or Creation, but these are the big pieces of the puzzle. After you read about Fossils, Irreducible

Complexity, Chance, Thermodynamics, Isotopes, the Old Earth and Noah's Flood, you will find 23 questions and answers about other puzzle pieces in nature too. You can read any of these sections and understand the material without having to read any other section; each answer is designed to be complete. The puzzle pieces are there for anyone to see and, in fact, the Bible says, "*The heavens declare the glory of God...*" so let's go take a look at the big picture.

How to remember this:

Evolution is only science F.I.C.T.I.O.N.

F.I.C.T.I.O.N.

- **F** = FOSSILS
- **I** = IRREDUCIBLE COMPLEXITY
- **C** = CHANCE
- **T** = THERMODYNAMICS
- **I** = ISOTOPE
- **O** = OLD EARTH
- **N** = NOAH'S FLOOD

Chapter 7

F = Fossils: What do they tell us?

Fossils are found mostly in sedimentary rocks. They are formed when the hard parts of a living thing, like bones, teeth, or wood, dissolve away and are replaced by minerals. In other words, the hard parts are literally turned into stone, or petrified. Scientists don't really know how this process takes place in nature, but many fossils are very well made. Some fossils of wood, for example, still show tree rings and the color of the original wood in the rock.

Of course, everyone knows that when something dies, it rots. Bacteria and scavenger creatures will eat everything unless the body is buried very fast. In water, dead fish and other creatures may float to the top and be eaten by scavenger birds, other fish and bacteria. The chances are very small that animals in the water or on land will be naturally buried quickly so they may become fossils. Evolutionists know this is true, so they believe fossils should be very rare.

The truth is, however, that fossils are not rare. They are, in fact, packed by the millions into huge areas. There are places where there are miles of bones, all fossilized. Most fossils are shellfish, like clams, but there are places where there are so many land-animal fossils buried in the same area that they can't be measured. One place, called the Karoo formation, has about 8 billion fossil skeletons![10]

The Bible teaches that there was a flood that covered the whole planet. This event probably covered the land areas of the Earth in huge waves. Each wave would have mixed up dirt, rocks, dead animals and plants and covered many dead creatures suddenly with mud. Heavier or special shaped rocks and plants would have settled to the bottom of each layer. Creatures that normally would have rotted away quickly could have been preserved as fossils. If this is what happened, we should see millions of well-preserved fossils all sorted out in different layers of sedimentary rock – which is exactly what the world looks like today.

But there's more. If life came about by Evolution, the fossils of creatures somewhere in-between one kind of creature and another should be there for us to find. This should be true because Evolution is supposed to happen when tiny changes, which we call "mutations", are passed on from parent to child. These mutations, after many years, are supposed to build up in a group of creatures until the entire group has changed into a totally different creature!

The in-between forms Evolutionists expect to find are called "*transitional forms*" and there should be millions upon millions of these creatures in the fossils. On the other hand, if God created animals and plants as basic kinds, we should see many variations but we should not see any clear examples of one kind of creature slowly changing into another creature. Either way, the fossils can tell us the real story.

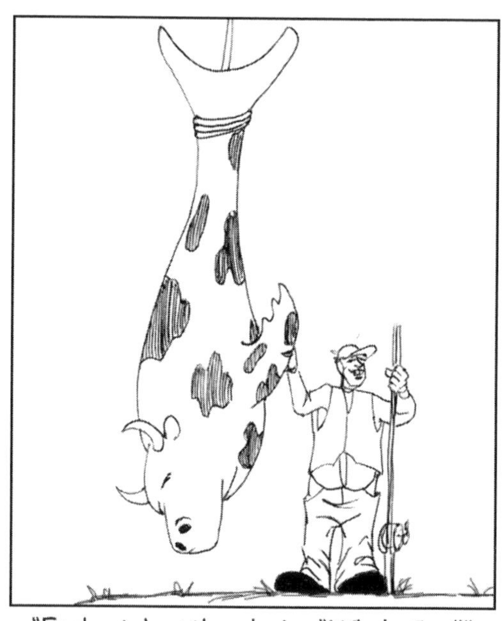

"Earl catches the elusive "Whale Cow!!"

As an example, think about bats. Bats are not birds. They are winged mammals that can fly in total darkness using an echo-listening system to find their way. Bats also have special wings that are actually made of skin membranes between their very long fingers.

Evolutionists believe that millions of years ago, a group of rodents, like rats or hedgehogs, began having offspring with longer and longer fingers. This means that by accident these rodents got brand new information in their DNA code. Gradually, some of these long-fingered rats gave birth to offspring with skin stretched between their super long fingers. After many years, just by luck, a group of long-fingered rats were born that just happened to have the perfect muscles needed to use their web-shaped hands to fly – in the dark. No one knows how these mutant rats could have survived but Evolutionists have to believe they did.

But that's not all. A bat has such good hearing they can hear the echo of their own voices over the voices of other bats. It works like this: bats make a chirping sound that bounces off an object. The bats listen for the echo of the chirp and can tell when they are too close to something. This is similar to the sonar systems ships use to find out the depth of water. This sonar-system is so good a bat can fly in complete darkness with thousands of other bats and not run into anything. Instead, they "*see*" with their ears.

No one can imagine how bats could have learned to see in the dark by chirping and listening for an echo. No one can figure out how the ancestors of today's bats could have survived with super long fingers that didn't yet work as wings. No one knows how these strange mutants could have given birth to offspring with just the right muscles and just the right brain to fly in the dark. No one can imagine how just the right new information was accidentally put into their DNA code so that everything would work perfectly.

Clearly, after so many years, we should be able to find many in-between or transitional forms between rats or some rat-like creature and bats. Evolutionists have looked for a long time, but all they ever find are 100% bat fossils and 100% rodent fossils. They have never found any transitional forms for bats, even though they have found bat fossils that are supposed to be about 50 million years old.[11] That's amazing.

This lack of in-between forms is found for every major kind of plant and animal. The fact is, no one has ever found any "for certain" transitional form between any two major animal kinds. The only creatures Evolutionists say might be transitional forms are not convincing to many scientists – even Evolutionists.

A good example would be Archaeopteryx (ark – ee – op – ter – ix). For years, science textbooks showed Archaeopteryx as a transitional form between reptiles and birds because Archaeopteryx has a long tail, claws on its wings and teeth in its bill, unlike living birds today. But after a few years, it was shown that "*Archie's*" feathers are identical to modern birds, its wings are completely bird-like and it even has a wishbone, also called a furcula, which is a bone found only in birds. In addition, there are three living birds today that have claws on their wings – the Hoatzin, the Touraco, and the Ostrich. Finally, "*Archie's*" teeth don't prove anything either because there are many animals that have characteristics of other animals that are not considered transitional forms.[12] Think about the platypus, a mammal that lays eggs like a bird. Nobody thinks the platypus is evolving into a bird just because it lays eggs![13]

Every creature Evolutionists think is a transitional form turns out not to be after some careful study. There are millions upon millions of fossils in museums today, so there should be many transitional forms for us to see. The truth is, there are no convincing examples of transitional forms between any of the major kinds of plants or animals.

It is also true that fossils are not rare.[14] It looks as if every kind of creature was complete right from the start and was suddenly buried so they could become fossils – just like the story of the Creation and the Flood we read about in the Bible.

How to remember this:

Evolution is just a long NIFTY tale, but without in-between forms it's one epic fail.

N.I.F.T.Y.

NO IN-BETWEEN FORMS TO YAK about!

Chapter 8

I = Irreducible Complexity: Why things don't work unless all the parts are in place!

Imagine walking along a beach one day when you discover a small, round, gold object on the sand. You pick it up to discover the object is flat. One side has tiny numbers and two pointers on it, one smaller than the other. You look closer and realize the pointers move around the numbers and there is a ticking sound coming from the object. Finally, you notice that the numbers seem to match the time of day.

Your first thought might be that millions of years of accidental shifting of sand, water and wind brought all of these pieces together to create this useful object. You might imagine a piece of metal being pushed along by a stream until it bumps into two tiny sticks. As you think about these things, however, you notice that one side of the object can come loose. When you take off the loose side, you find dozens of complicated gears, springs and coils. This complicated machinery drives the pointers. Suddenly, you're imagining that this object could have come about by the accidental shifting of sand, water and wind seems pretty silly. Obviously, all of these parts had to be put together or the object would not work.

It does seem obvious that a watch was planned and made by someone rather than an accident of sand, water and wind. A watch is an example of something that has many complicated, well-matched parts. Each part of the watch needs the other parts in order to work. If one of the parts is missing or doesn't do its job right, the whole watch doesn't work. This is called "*irreducible complexity*".[15]

A single animal or plant is made up of thousands of complicated parts. Each of these parts is also made up of thousands of even smaller complicated parts. Evolutionists believe that all of these parts came about and were put into the exact right place by a natural step-by-step process. This is like imagining that a watch was built by the natural shifting of sand, water and wind. This natural step-by-step process is supposed to happen when mistakes, also called mutations, in the DNA code happen when animals or plants reproduce – all by accident. Offspring with the new DNA code information are changed by this mistake. This tiny, accidental difference somehow helps the new creature survive better than the parents. After many millions of years, these tiny changes are supposed to change one kind of creature into another. Remember, these changes are supposed to make complicated, well-matched parts. Each part needs all the other parts in order to work. If one of the parts is missing or doesn't do its job right, then the whole thing doesn't work.

It is true that all kinds of amazing things can happen by accident. But can mistakes from one generation to another explain things that are irreducibly complex? The natural world is filled with hundreds of examples of amazing creatures and living systems that are irreducibly complex.

For example, think about butterflies. Every elementary student knows that caterpillars become butterflies. But caterpillars are very different from butterflies. Caterpillars are designed to crawl. They have mouthparts designed for chewing and eating leaves. They have a stomach system designed to digest leaves and they have special glands designed to spin silk. The interesting thing about caterpillars is that they have no reproductive organs. In other words, they don't lay eggs or have baby caterpillars.

Instead, as we all know, caterpillars spin a cocoon and later come out as a butterfly. Butterflies, however, are designed to fly, not crawl. They do not have mouthparts designed for chewing and eating leaves. Instead, butterflies have a long, tubular mouthpart designed for sipping nectar out of flowers. They have a stomach system designed to digest nectar instead of leaves and they have no special glands designed to spin silk. The interesting thing about butterflies is that they do have reproductive organs and can lay eggs - but butterfly eggs only hatch into caterpillars.

These creatures seem to be completely different from each other but they are, in fact, the same animal. If Evolution is true, somehow a caterpillar-like creature slowly changed, just by accident, into a butterfly. But how was that supposed to happen when caterpillars do not have any reproductive organs? Perhaps we can imagine a butterfly-like creature that slowly changed, just by accident, into a caterpillar. But why would the new caterpillar-like creature be perfectly designed to turn back into a butterfly?

Even if we could imagine a caterpillar-like animal that did have reproductive organs, what about explaining the amazing cocoon system? A caterpillar in a cocoon actually melts into a mix of different cells. Different types of butterflies re-arranged this soup in different ways too – they are all different.[16] Scientists have no idea how butterflies are built from this mix. They have no idea how the building process is directed either.[17] This is an example of an irreducibly complex system. All of the parts have to be in place from the beginning, or the system just doesn't work. There is no natural, step-by-step way to explain how this kind of system started in the first place.

Some scientists have some very exciting imaginations but they have not shown any examples from the past that caterpillars were ever anything other than caterpillars or that butterflies were ever anything other than butterflies. In fact, Evolutionists cannot explain where the entire group of animals called insects (arthropods) came from in the first place![18] All we find in the fossils and from examples of these creatures today is what every elementary student studies in science class – caterpillars turn into butterflies. It appears these creatures had to have been planned and built. Their life-system had to have been complete, right from the start, just the way the Bible says. There is no way anyone can show how these animals came to be by any step-by-step process.[19]

There are thousands of examples in nature that are irreducibly complex.[20] In fact, there are so many examples, it seems easier to believe a pocket watch could be built by accident from the movement of the wind, sand and sea. On the other hand, it seems much more reasonable to believe a pocket watch was designed and built for a purpose, so it also seems more reasonable to believe the same thing about living things. After all, living things like butterflies are by far more complicated than a pocket watch anyway.[21]

How to remember this:

Butterfly hearts come from caterpillar PARTS.

P.A.R.T.S.

PARTS ARRANGED RIGHT TEARS-down any STEP-by-Step explanation.

Chapter 9

C = Chance: What are the odds?

Imagine that you have a choice between watching one of two movies. You only have time to see one of the movies but you want to see both, and both are good choices. In the end, you decide to flip a coin to make your decision. What are the chances you will see movie "A" or movie "*B*"? Imagine that you use a standard coin and decide that the "heads" side will represent movie A and the "*tails*" side will represent movie B. In this case, the chances, also called "*the odds*", of getting A or B is one chance in two. This is because you only have two choices. No matter how many times you flip the coin, the chances you will get A or B will be one chance in two on every toss.

Mortimer tries to even the odds!

People can understand the chances of something happening by figuring out the math. For example, think about 10 flashcards numbered from one to ten. If I shuffle the cards at random and then lay them out on a table in a line, what are the chances that the cards will end up coming out in order from one to ten? The answer is one chance in 3,628,800. In other words, there are 3,628,800 ways the flashcards can be put together - but only one of those ways can be in order from one to ten. Incredible, isn't it?[22]

Understanding the chances of something happening can be important to know. Most of us do not spend a lot of time worrying about being struck by lightning. This is because the chances any one of us will be hit are very small. Most of us don't even know what these chances really are, but we know the chances are so low, we don't worry about it. Actually, in America, the odds are about one chance in 750,000 that you will be struck by lightning in any given year.[23]

Evolutionists believe that all life came about by chance. If Evolution is really what happened, the odds life could come about by chance need to be pretty good. If we can figure out these odds, they will tell us a great deal about how likely Evolution is to be true. In order to think about these odds, we have to look at the basic building blocks of life. We need to see what the chances are these building blocks could be put in just the right order for life - by chance.

All life as we know it is based on protein molecules. Proteins are, in turn, made up of long chains of even smaller molecules called "amino acids." The amino acids are all arranged in a very specific pattern. This pattern of amino acids allows proteins to do specific jobs. Living things use these proteins for many things, including building body parts and digesting food. Proteins have specific shapes so they can fit into other proteins, like keys in a lock, in order to make things work in living things. It is a lot like the gears, wheels and springs in a watch. If just one of the gears is the wrong shape, the whole thing won't work.

Just like the gears in a watch, proteins are folded into just the right shape too. This means that only ONE pattern of amino acids will make them fit. Think about our flashcards again for a moment. There are 3,628,800 ways that flashcards numbered one to ten can be laid out on a table, but what if only one of those ways works for our needs?

Think of it this way. Let's say that laying out our flashcards from one to ten is called "*pattern* A". Now, imagine another group of flashcards laid out a different way. Let's call this second group of flashcards "*pattern B*" and pattern B is absolutely needed to do a very, very important job. But Pattern B won't work unless Pattern A is there first!

Even worse, what if the chances pattern B will exist in the right shape to fit pattern A are also one chance in 3,628,800? What are the chances we could lay out pattern A *and* pattern B in just the right order for the whole system to fit together? The odds are so huge it would be like the chances a hurricane blowing through a junkyard could build a brand new, perfectly built, 747 jetliner (an airplane with over 6,000,000 interrelated working parts).[24] That doesn't happen every day and it seems very unlikely it ever will happen.

Remember that proteins are chains of amino acids. These amino acids are like our flashcards. Scientists trying to figure out what the chances are that proteins could be put together in just the right order have done all kinds of studies. These studies left scientists stunned because the odds are so small they are just plain impossible. One study took a look at one type of protein called a cytochrome C. Even if all the parts needed for this protein were in place, there is only one chance in 2×10^{-94} that all the parts could be put together the right way by chance. To understand this number, think about making 100 flashcards to represent all of the parts needed to make cytochrome C. Now imagine that we had 100 billion people and each of them had a set of these flashcards (a billion is 1,000 millions). If all 100 billion people laid out their cards one trillion times every second (a trillion is 1,000 billions) for

20 billion years (20,000 millions), you still wouldn't have enough tries to get the one right pattern for a cytochrome C.[25]

The building blocks of life, such as proteins and amino acids are by far more complicated than the world's most expensive watch. Even some well-known and respected Evolutionists admit that the most basic form of life could not have started by chance.[26] These odds tell us life had to have been designed and built. There is simply no other explanation that makes sense.[27]

How to remember this:

> The odds against Evolution are really so high, it would be like believing tornados build JETS for the sky.

J.E.T.S.

> **J**UST one pattern for proteins and **E**XTREMELY bad odds makes it **T**OTALLY impossible life **S**TARTED by chance.

Chapter 10

T = Thermodynamics: Why everything falls apart!

Imagine taking a brand-new car and parking it in a field. If you left the car alone for a million years, what result would you expect to see? No one would really expect to see the car change into something more complicated, like an airplane. Instead, we would expect the car to rust away into nothing. You wouldn't have to do anything to the car at all – it will simply break down all by itself.

We can see this idea everywhere we look. It takes work to clean up a house but left alone, the house will become a mess again. Everywhere we look in nature, we see things that start out in a complicated order slowly becoming less complicated and less orderly over time. Scientists call this process "*the Second Law of Thermodynamics*".[28] There is more scientific evidence for the Second Law of Thermodynamics than for any other scientific law.[29] All anyone has ever seen in nature is that complicated things break down into less complicated things over time.

At first glance, there seems to be some exceptions to this rule. For example, a seed is less complicated than a tree and a baby is less complicated than an adult. But this isn't really an exception to the Second Law for three reasons: 1) a seed and a baby each have a set of instructions in their cells that guide the growing process. This means they each have guiding information. 2) A seed and a baby each have a means, or a motor, for using energy to grow according to their growth plan. A seed uses photosynthesis and a baby uses digestion of food, among other things. 3) Once the seed and the baby are finished with the growth plan, they will eventually die. In other words, only planned, intelligent energy can do the work needed to get around the Second Law but this only works for a short while.[30] Once the plan and the energy are gone, the Second Law takes over once again.[31]

The growth plan in a plant or animal is stored in its DNA code. DNA is like a super computer. It has the growth plan for the plant or animal and it tells the cells of the creature how and when to grow. When an animal or plant reproduces, the DNA plan is passed on to the next generation but every once in a while, a mistake, called a "*mutation*", happens. Evolution is the idea that these mistakes can cause the new generation to survive better than the parents. Evolutionists think that over millions of years, these mistakes can cause animals and plants to become more and more complicated.

It's easy to see that the idea of Evolution goes against the Second Law. It is also easy to see that living things are much more complicated than non-living things. If Evolution is what really happened, then very complicated living things came from less complicated non-living things – all by accident. But this idea breaks the best proved of all scientific laws.[32] Of course, if a scientific idea doesn't obey known laws, then there is a serious problem with the idea.

Some scientists think that energy from the sun caused living things to become more complex. They think that adding enough energy to a system will get around the Second Law. But the sun's energy is not intelligent or ordered energy. Blasting a system with the sun's rays only breaks the system down even more. Just having enough energy doesn't solve the Second Law problem.[33]

The Second Law of Thermodynamics is a big problem for scientists who believe that life came about by accident from non-living things. It is also a problem for scientists who believe that living things became more and more complex. But the Second Law is not a problem for people who believe God created the heavens and the Earth. In fact, it is just what we would expect to see.

How to remember this:

Just like DYNamite can blow up things like ceramics, so Evolution is blown up by Thermodynamics.

D.Y.N.

- **D** = DYNAMITE blows things up, just as everything in nature is breaking down, all the time.
- **Y** = YOU are able to think and only intelligent energy can get past the Second Law.
- **N** = NATURAL energy, such as the rays of the sun, does not provide intelligent energy – so Evolution breaks the Second Law.

Chapter 11

I = Isotopes: How geologists date rocks!

Evolution is the idea that mistakes in DNA, passed on from parent to offspring, can cause the new generation to have an ability to survive better than the parents. Evolutionists think that over millions of years, these mistakes can cause animals and plants to become more and more complicated. Eventually, we are told, these tiny changes will cause one kind of animal or plant to change into a completely different kind of creature.

If this is true, we should see mistakes in DNA that cause a new generation to gain some sort of benefit. The truth is, however, that most of these mistakes, also called mutations, are harmful or even fatal.[34] The tiny number of mutations that are not harmful usually have no effect on the creature at all. It would seem that the chances of a good mutation helping an organism survive better will be very small.[35]

Since mutations are so rare, Evolutionists think it must have taken millions of years for the "*good*" mutations to happen. If the Earth is not very old, there simply isn't enough time for "*good*" mutations to show up. This is one of the main reasons these scientists believe the Earth is 4.5 billion years old.

On the other hand, God could have made the world millions of years ago or just thousands of years ago. The Earth does not have to be any one age for Creation but it absolutely *must* be billions of years old if Evolution is true.[36] Either very old or very young, Creationists expect to see living things designed to change in small ways to help them survive, but these small changes will never change one kind into a completely different kind.

One of the most important reasons many scientist believe the Earth is billions of years old comes from radioactive dating of rocks. Evolutionists started experiments with this method in the 1930s and they have used these experiments to convince people the Earth is old enough for Evolution to be possible. They have been very successful too since most textbooks and scientific papers teach that radioactive dating "*proves*" the Earth is about 4.5 billion years old.

There are some very serious reasons that radioactive dating doesn't really tell us the actual age of a rock. Most people don't know about these reasons but even many Evolutionists realize that radioactive dating is not very dependable.[37]

In order to understand how radioactive dating works, it is important to understand a little about atoms first. An atom is made of protons, neutrons and electrons. Protons have a positive electrical charge, electrons have a negative charge and neutrons have no charge at all. An atom that has the same number of protons but a different number of neutrons is called an isotope. Some isotopes can be very unstable and may give off energy. Scientists call this energy "*radioactivity*".

Radioactive minerals may be found in rock that was once so hot it was molten lava. Rocks formed in this way are called *igneous rocks*, and many of them have radioactive minerals, such as Uranium, caught inside. Radioactive minerals have unstable atoms and can give off a large amount of energy. For example, the atoms in Uranium 238 cannot hold onto their energy permanently, so they lose energy slowly over the years. It is as if little pieces of Uranium 238 are regularly falling off. Scientists call this process *radioactive decay*.

Other minerals, such as Lead 206, have proton and neutron arrangements that look similar to Uranium 238. Scientists think that after many years, radioactive elements like Uranium 238 will release so much energy that they will change into stable elements like Lead 206. This is why scientists call the stable atoms that are supposed to come from radioactive minerals "*daughter elements*".

We can measure how fast radioactive minerals lose energy. Many scientists think the speed of this decay has always been the same, so they just add up how long it should take for the radioactive mineral to become a non-radioactive mineral. For example, half the atoms in a sample of Uranium 238 will probably change into Lead 206 in 4.5 billion years. This is called the half life of Uranium 238. Scientists try to date rocks by measuring how much radioactive material is in a rock and comparing it with how much daughter material is in the same rock.

There are a number of problems with this idea. For example, if we were looking at a sample of rock that has Uranium 238 and Lead 206, how can we know for certain there was only Uranium in the rock when it first formed? If there was some Lead 206 in the rock when it was first made, we might think the Lead came from the Uranium. If we just assume all the Lead came from Uranium decay, then the rock will appear to be very old. The truth is, there is no way to tell how much Lead in the sample came from Uranium and how much Lead was there from the start.

The second problem is that no one can say for sure Uranium has always broken down into Lead at the same speed. We have been measuring the speed of Uranium decay for about 100 years now, but Uranium has been around much longer than that. It looks as if Uranium decay is fairly steady, but there is no way to be absolutely certain.

The third problem is that our rock sample may have been polluted somehow in the past. What if some Lead or Uranium was washed into the sample by ground water, or what if some Lead was mixed into the molten rock just before it became solid? Some scientists think that pollution of rock samples is such a serious problem that radioactive measurements of rocks may have nothing to do with their ages.[38]

Lastly, scientists have done radioactive dating on rocks of a known age and come up with dates that were wrong by millions of years. They have even found different ages in the same rock. For example, there is a volcano in Hawaii called Hualalai that erupted in 1801. Scientists tried to date this lava 12 times and came up with dates anywhere from 140 million years all the way up to 2.96 billion years.[39] It seems that radioactive dating doesn't really prove anything at all.[40]

How to remember this stuff:

Radioactive dating is a kind of a clock, but it can't really tell us the age of a ROCK.

R.O.C.K.

- **R** = RADIOACTIVE minerals may not be the only minerals in the rock when it first formed.
- **O** = OUTSIDE sources may have contaminated the rock.
- **C** = CONSTANT decay rates may not be constant after all.
- **K** = KNOWN ages for some rocks have radioactive dates that are off by millions of years.

Chapter 12

O = Old Earth: Don't judge the book by its cover!

The idea of Evolution is based on mistakes in the DNA code of life-forms. These mistakes, also called mutations, are believed to cause one kind of living thing to transform into another over time. No one has ever seen a truly good mutation. No one has ever seen one basic kind of living thing change into another either so Evolutionists think this process is so slow no one can live long enough to see it happen. If this is true, the Earth must be very old and as a result many Evolutionists believe the Earth is at least 4.5 billion years old.[41]

The Bible teaches that God created everything using His own power. Some people who believe the Bible think the Creation happened 7 or 10 thousand years ago. Others who believe the Bible think the Creation may have happened millions of years ago. A younger Earth fits the Bible's story of creation much more naturally than an old Earth idea. The truth is the Earth does not have to be very young or very old for Creation science. Creationists can discuss different ideas about an older or a younger Earth but Evolutionists have no choice. The Earth absolutely must be billions of years old if Evolution is true. This means that any evidence possibly showing a younger age could be deadly to the idea of Evolution.[42]

For centuries, people believed the Earth looked as it does now as a result of Noah's world-wide flood but in 1785, James Hutton wrote a book that challenged this idea. In 1832, Charles Lyell expanded on some of Hutton's ideas. In these books, Hutton and Lyell argued that landforms like canyons and valleys took millions of years of wind and water erosion to form. They believed the same slow processes of nature we see all around us today have been operating at basically the same speed over time. This idea is called *uniformitarianism* and is best known by the phrase, "*the present is the key to the past*". This is one of the main reasons some scientists think the Earth is billions of years old.

Today, however, many scientists are beginning to accept the idea that landforms could have been formed very rapidly from time to time. After all, they know the occasional catastrophe such as an erupting volcano or a large flood scouring the ground can make major changes. Still, most scientists believe these catastrophes only happen occasionally over billions of years. This idea is called *episodicity*. Scientists are beginning to think this way because most major landforms cannot be explained just by today's weather and erosion patterns.

For example, the Hudson River Canyon is a vast cut into the Earth with walls nearly three quarters of a mile high - but it is 7,000 feet underwater! The floor of the canyon is as much as 3,600 feet deep from canyon rim to floor. It covers an area the size of the state of Connecticut making it nearly as large as the Grand Canyon. Modern ideas about how this canyon could have formed don't add up.[43]

Another canyon that modern scientists cannot explain is the Bahama Canyon in the Bahamas. This enormous canyon is 14,000 feet deep, 40 miles wide and 125 miles long, more than twice as big as the Grand Canyon in Arizona.[44] There is also the Columbia-Plateau, a flow of lava that covers 200,000 square miles. There are fossil beds where millions upon millions of fish were buried so fast they didn't have time to decay. Evolutionists cannot explain how processes operating so slowly today could have made these things.[45]

One explanation that Creationists consider is the world-wide flood of Noah's days. Such a mighty flood would have caused tidal waves that scoured the continents over and over, laying down massive layers of mud and ripping out huge canyons as the water pulled away. There are so many places where landforms absolutely had to have been formed quickly that some scientists have lost count![46]

There are other things that show a younger date for the Earth than just landforms, so many, in fact, that Creationists have written dozens of books to point them out. One of these facts is the amount of helium in the Earth's atmosphere. Radiogenic helium comes from radioactive decay in the Earth's soil. We can measure how much radiogenic helium comes out of the ground and how much there is in the air. The amount of radiogenic helium in the air today would have been built up in less than two million years.[47]

Another fact that shows a younger age for the Earth is the saltiness of the ocean. Many minerals, such as magnesium, uranium, and nickel, besides salt, are dissolved into the sea each year. Evolutionists have figured out how long it would take for the oceans to get as salty as they are today if they were as fresh as drinking water in the beginning. All of the minerals would have been built up in much less than two million years. In fact, it would only take 50,000 years to get the amount of copper we have in the oceans today and 18,000 years for the amount of nickel. No one knows how much of these minerals were in the first ocean, and no one knows whether the dissolving rates were different in the past. The only thing we know for certain is that every year, some of these minerals are washed into the ocean. Even if this amount is very, very small, the oceans can't be more than two million years old.[48]

There are many more processes that give younger dates for the Earth than processes which give older dates.[49] We can't be certain exactly how old the Earth really is but we can be certain it cannot be *as old* as Evolutionists need it to be to try and support their model.

Creationists will continue to discuss facts for one age or another and they can agree to disagree, but Evolutionists have no choice – they have to try and explain the vast number of facts that point to a younger age. So far, the best explanation is that the Earth is simply not that old.

How to remember this stuff:

> The Earth could be young or it could be old, but it's not old ENUF for Evolution to unfold.

E.N.U.F.

- **E** = EVOLUTION requires a great age for the Earth.
- **N** = NOAH'S flood built land-forms as we see them today.
- **U** = UNIFORM rates for land-form building must have been much greater in the past.
- **F** = FORMATIONS such as the Hudson River Undersea Canyon can't be explained by present day processes or rates of erosion.

Chapter 13

N = Noah's Flood: Was there really an Ark, a Flood and a guy named Noah?

Many scientists and teachers don't take the story of Noah and the Flood seriously. Some Evolutionists call the story a fairy tale and consider anyone "*unscientific*" who believes that the Flood actually happened.[50] Instead, they believe all land-forms that exist now can be explained by the same slow and gradual chipping away of wind and water that we see today, with the help of the occasional catastrophe. They also teach that all life-forms today can be explained by slow accidental mistakes in DNA between one generation and another. Most Evolutionists and textbooks picture the story of Noah and the Flood as a myth.

But the Bible says that some time after God created the world, humans became so bad they needed to be destroyed. God told Noah to build an ark, which is something like a large barge, in which to keep alive two of every kind of land-dwelling, air-breathing animal. Noah took his wife, his three sons and their wives into the ark with him, along with the animals. Together, they were the only land creatures and humans to survive the Great Flood (Genesis, Chapter 7). Plants, marine animals and most insects survived outside the ark and all living things we see today are the descendants of these survivors.

Christians, for the most part, believed the story of the Flood was an actual event in history for centuries. This changed in the 1830s when people began to think that the same rates of erosion we see around us in the present built all the land-forms we see today over long periods of time. Before long, many church leaders began writing about the Flood as if it were just a spiritual story instead of real history. Others began to teach that the Flood was not a world-wide event but only a local flood that was exaggerated in the Bible after many years of re-telling the story. These ideas became very popular and even today, not all Creationists believe the Flood story actually happened and neither do all Christians or church leaders.

The Bible, however, gives at least seven reasons to think the Flood was a real event and that it covered the entire planet. First, the Flood was deep enough for the ark to float over the tallest mountains. Second, the Flood lasted for 371 days, longer than any known weather disaster today. Third, the Bible describes "*the breaking up of the great deep*" (Genesis 7:11) which seems to point to major earthquakes and volcanic eruptions world-wide. Fourth, the ark is described as the largest vessel ever built by man until 1858, too big for the needs of a local Flood. Fifth, the Bible says the purpose of the Flood was to destroy a world-wide population of humans. Sixth, Noah's story is similar to Flood stories and legends from nearly every culture on Earth.

But most importantly, other writers in the New Testament of the Bible quote the book of Genesis 165 times. One hundred of these quotes are from the first 11 chapters of Genesis. Jesus Himself quotes from Genesis 25 times and nowhere is there any hint these people thought of Genesis as anything other than real history.[51]

What about all the animals?

Many people find the idea of Noah's Ark impossible to believe. How could anyone cram millions of animals on a boat, especially huge animals like elephants or vicious creatures like dinosaurs? But a closer look at Noah's story paints a very real picture.

In the first place, the Bible says Noah's Ark was gigantic: nearly 450 feet long, 75 feet wide, and 45 feet high! The Ark was at least as long as a football field. It also had three decks inside, giving it a total inside volume of 1,396,000 cubic feet and weighing approximately 13,960 tons. This would be enough room to store as much as eight freight trains pulling 65 railroad boxcars each. That's 520 boxcars.[52]

The number of animals sent to Noah wasn't nearly as large as most people think, either. Noah had at least two of every kind of land-dwelling, air-breathing creature. He also took seven individuals of certain special animal groups that were considered important for ceremonial purposes. This means Noah did not have to deal with water-dwelling animals.

He probably didn't have to deal with most insects, either. Many of them do not breathe air the way you and I do. Many insects and their kinds could have survived as larvae on floating plant material or emerged later from eggs. If Noah took some insects on the ark, they certainly didn't take up much room. Noah had to be concerned about mammals, birds, reptiles, some amphibians and some insects.

There is also a difference between a "*kind*" and a "*species*". A *kind* is a much more general term, like the horse-kind or the dog-kind. All the various breeds of these animals can come from a single breeding pair. Noah didn't have to take every breed of dog or cat, only the type or kind that God knew had the ability to give birth to a wide variety.[53]

A kind can produce many varieties, but it will never change into a different kind. For example, we can get more than 500 varieties of sweet pea from just one plant but all of these varieties are still sweet peas.[54] Dogs are still dogs even though there are Great Danes and poodles, so Noah only took the basic kind of each type to re-populate the Earth.

After many years of study, Creationists believe the "*kind*" is probably the same as the scientific term "*family*", so there may have been as few as 2,000 animals on the ark.[55] To be generous in our thinking, we can estimate Noah needed to take between 16,000 and 35,000 animals and this number includes all animals we know today only as fossils, such as dinosaurs.[56]

Noah would have taken young, healthy animals instead of adults too, so, on average, the animals on the ark would have been about the size of a rat. In fact, less than 11% of these animals (that would be between 1,760 to 3,850 animals) would have been much larger than a sheep.[57]

We can carry well over 50,000 sheep on eight freight trains pulling 65 boxcars each. In fact, there was enough room on the ark to fit 35,000 animals into only half the available space. There was more than enough room to fit in all the animals and food to feed them and still have empty space![58]

People have studied large farms where people manage to feed thousands of animals and take care of their waste products. It is very possible for just a few people to care for tens of thousands of animals. For example, Noah could have built enclosures with ramps to take away manure. He may have put the manure in long pits between rows of cages where insects and worms would have consumed much of the waste. The insects and worms could have been used to feed some animals.[59]

It is also possible the ark had a large "*moon pool*", an opening allowing water from the outside to move up and down into a large inside pool. The movement of water in the moon pool could have kept air flowing through the Ark and taken away much of the waste from the manure pits that was not consumed by insects and worms. Many animals in a darker environment with a raging storm outside may have hibernated too. It is very possible, using simple ideas like ramped cages and a moon pool, for eight people to have taken care of 35,000 animals for 371 days.[60]

What about Dinosaurs?

It is hard to imagine Noah taking something as dangerous as a *Tyrannosaurus* or some other large, meat-eating animal on the ark. Many Creationists believe that dinosaurs and humans must have lived at the same time prior to Noah's Flood, so Noah would have put them on the ark also. This looks like a big problem when you think about how large dinosaurs grow and how frightening they must have been.

The answer to this is that young dinosaurs were probably not much larger than large lizards. In the 1920s, paleontologists discovered fossilized dinosaur eggs. These eggs let us know that dinosaurs started out very small. They may have been like modern crocodiles and other reptiles that never quit growing in life. Crocodiles can grow to 50 times their birth size. Dinosaurs just hatched from eggs were probably far easier to manage than the full-grown models. For example, you can pick up a baby crocodile (very carefully), but you shouldn't try to pick up a full-grown crocodile unless you want to get eaten.

The truth is we really don't know much about how dinosaurs actually behaved anyway. All we have is their bones, teeth, and some footprints. Their teeth look terrifying, but they don't tell us how dinosaurs acted in life. For example, the mighty *Tyrannosaurus rex* had thick, six inch teeth but some scientists now think these teeth were rooted too shallow in *rex's* jaw to chew raw flesh and bone. Tyrannosaurus just might have been a scavenger like a vulture, eating only dead things. No one knows for sure, because we have less than twenty Tyrannosaurus skeletons to look at anyway.

Myth busters

Before the Flood, the world was very different. There was a warm climate, higher air pressure and a shield of water vapor in the atmosphere that kept out harmful radiation from the sun. It was God Himself that caused the right animals to come to Noah in a massive migration and it is very possible that God allowed a comet or a meteor to hit the Earth causing the Flood. Thinking about all of these things makes Noah's story seem much more realistic.

But ideas about exactly how many animals Noah would have had to take and moon pools do not prove that Noah really did have the largest floating zoo in history. They do, however, help explain how this story could have happened. In the end, you will have to decide for yourself if you want to believe the Bible or not.

How to remember this:

Genesis tell us about the world's greatest zoo, it tells us the animals came to NOAH, TOO by TOO.

N.O.A.H., TOO by TOO

- **N** = N<u>o</u> **porpoises** for Noah's purposes, land animals and maybe, they all came as babies.
- **O** = ONLY *kinds* were involved, today's species are solved.
- **A** = ARK was so large, fit 50,000 sheep on a barge.
- **H** = HUMAN beings, eight in all, could have cared for them all.
- TOO deep by TOO wide, means the flood was world-wide.

Question 1

What is the Big Bang? How do some people think everything got started in the first place?

The Big Bang theory is the idea most Evolutionists use today to describe how they think the universe got started.[61] According to this idea, all of the rocks, gases, dust, and energy in the universe, which scientists call mass/energy, was originally lumped together into one little ball. There was so much energy and weight in this ball that is was super-hot and super-compact. Some scientists say the ball could have been as tiny as an atom! For some reason that no one can explain, the ball exploded with a great big "*bang*" about 15 to 30 billion years ago.

After the ball exploded, streams of matter and energy spread out all over space. As it spread out, it began to cool and form the simplest element in the universe: hydrogen gas. Over time, Evolutionists believe, everything in the universe came from the Big Bang. This includes stars, planets and life on Earth, too.

There are some very serious problems with this idea. For example, no one knows where the super-hot ball of mass/energy came from in the first place. The only ways scientists can try to answer this question are not scientific. In other words, the Big Bang idea says that first there was nothing, then there was everything. No one knows why or how everything could just pop into being from nothing.[62] How could everything come from nothing just by accident?

Second, even if we can imagine a super-hot ball of mass/energy that came out of nothing, no one knows any reason why it would have exploded. If there really had been such a ball of mass/energy, it would have had super-powerful gravity. The gravity would have been so powerful even light could not have escaped from it. If the gravity in the ball was so intense, no one can explain what might make it explode. How could all the mass/energy escape from such powerful gravity?

Another big problem with the Big Bang idea is that we do not see any new stars being created today. Astronomers do see stars explode into supernovas from time to time. They also see huge clouds of gas and dust in space, but no new stars seem to be forming. There are hundreds of billions of stars in the universe. If stars came from the Big Bang, we should be seeing new stars lighting up all the time - but we don't.

The amount of material in the universe, also called "*mass*", is far too small to fit the Big Bang idea.[63] There's just not enough "*stuff*" in space to fit the Big Bang model. Also, if the ball had exploded in the vacuum of space, then the universe should be very smooth. The mass from a Big Bang would have spread out evenly everywhere. But the universe is not smooth. Instead, it is lumpy and uneven.[64] This doesn't fit the Big Bang model.

Lastly, have you ever heard of a bomb going off and creating an airplane? Of course not! Explosions only create disorder. Our universe, however, is very orderly. It is also very complex. This is not what we usually get from an explosion.

There are so many problems with the Big Bang theory that many scientists are ready to reject it. Still, it remains the most popular theory. This is because other ideas about how the universe may have started accidentally do not fit the facts of science any better than the Big Bang model.[65] There is, however, one theory that fits the facts of science very well. It begins with the mighty power of God. God is outside of time, space and matter and God said "*Let there be light*", and it was so.

How to remember this:

God said "*let there be light*", and BANG, it was so and boy what a sight!

B.A.N.G.

- **B** = **a** BALL of mass/energy = super-massive gravity, so how could it explode?
- **A** = AMOUNTS of mass/energy = too little for the Big Bang model, so where is the missing mass?
- **N** = NEW stars are not forming today, so why aren't they forming?
- **G** = GROUPS and clusters of stars shouldn't be there, why isn't the universe smooth?

Question 2

Did Evolution Happen Very Fast? What is Punctuated Equilibrium?

Sherlock Holmes, the great detective in the stories by Sir Arthur Conan Doyle, said, "*Once you have eliminated the impossible, whatever is left, no matter how improbable, must be the truth*".

If Mr. Holmes were investigating a suspect for murder, he would look for evidence that could tell him what had happened in the past. If there was no evidence that the suspect was guilty, he would decide that someone or something else must have been guilty. A lack of evidence would only prove the suspect was NOT the murderer.

"The Butler Did It!"

Unfortunately, people may *want* to believe someone is guilty of a crime, even if the facts say he is innocent. In this case, they might convict an innocent man. Ideas we already have in our minds are called "*pre-conceived ideas*". Pre-conceived ideas have a way of twisting the facts. This is also true of science. This is why it is important for scientists to work hard against the effects of pre-conceived ideas.

As we have already seen, there is a gigantic lack of evidence in the fossil record for Evolution. Some scientists try to explain this by saying the right kind of fossils have just not been found yet. But there are a great many fossils to study and many are in very good condition. This presents a problem for Evolutionists. One way some scientists are trying to explain this lack of fossil evidence is called "*Punctuated Equilibrium*".

Punctuated Equilibrium could really be called "*fast*" Evolution. The idea is that Evolution happens so quickly, there is no time to leave enough fossils to study how it happened. If Evolution happened very fast, then we should expect very few examples of in-between life-forms, also called "*transitional forms*", in the fossils. This is, of course, exactly what we see in the fossils. There are no good examples of real transitional forms.

Punctuated Equilibrium, or "*fast*" Evolution, is the idea that normal animal and plant groups stay pretty much the same for millions of years. Somewhere along the line, this steady situation is broken or "*punctuated*" by very fast Evolution. This is supposed to happen when a small part of a group of life-forms is trapped in a new and stressful place. Since the new environment is so stressful, natural pressure causes very rapid changes in the life-forms. Since these changes happen so fast, there is simply not enough time for good fossils to form. In other words, since there is no fossil evidence, this must mean Evolution took place too fast to leave any evidence. But how can a lack of evidence be the evidence you are looking to find? How does that prove anything?

Most textbooks today explain Evolution as a slow and gradual process. Evolution is supposed to happen because of small, accidental mistakes (also called "*mutations*") in DNA. The idea is that one kind of life-form gives birth to a new generation that has a tiny difference. These tiny differences are supposed to build up over millions of years so that a creature like a cow will eventually turn into a whale.

There are many problems with the "*slow and gradual*" idea of Evolution. For example, no one has ever seen a mutation that is truly positive in the first place. Besides, natural mutations are rare. Since this is true, it seems very unlikely that tens of thousands of just the right kind of mutations could all happen very quickly.

Some Evolutionists think that Evolution may have happened so fast it would be as if a reptile once laid an egg and a bird hatched out of it. This was called the "*hopeful monster*" theory of Evolution. At the time it was first proposed, the idea was laughed at by both Evolutionists and Creationists. Over the years, however, this "*hopeful monster*" theory has become more popular. This is because the evidence for slow and gradual Evolution isn't shown in the fossils. Of course, you would have to have *two* hopeful monsters born at the same time - one male and one female. If there was only one hopeful monster, it wouldn't be able to pass on its new DNA.[66]

Not all scientists who believe in Punctuated Equilibrium believe in the hopeful monster theory. Still, no one has ever seen one kind of animal or plant change into a completely different form. Even after thousands of years of people breeding farm animals to try and get the biggest and best animals, cows are still cows and sheep are still sheep. Punctuated Equilibrium has never been seen and has no evidence to show that it ever happened.[67]

How to remember this:

Punctuated Equilibrium as science is sunk, without any evidence it's just really JUNK.

J.U.N.K.

- **J** = JUMP, a cow turns into a whale so fast there's no time to leave any evidence.
- **U** = USUALLY life-form groups stay basically the same.
- **N** = NEW environments supposedly cause fast Evolution.
- **K** = KEY fossil evidence that is lacking does NOT prove fast Evolution took place.

Question 3

The Razor's Edge: What is the universal constant?

Life is incredibly complex. It is also very delicate. If conditions are not just right, life cannot exist. Planet Earth just happens to have everything needed for life. For example, it is exactly the right distance from the sun. If it were any closer to the sun, all liquid water would boil away. If it were any farther away, all the water would freeze. Without liquid water, life cannot exist. We have over 350 million cubic miles of liquid water on Earth, but scientists cannot find one single drop of liquid water anywhere else in the universe.

Oxygen is another element we must have for life. But oxygen can also break things down. This is what happens when metals rust – they "*oxidize*". If you have too much oxygen, it will break down the molecules that living things need. But if you don't have enough, living things cannot survive. The air on Earth is about 21% oxygen, just enough for life but not too much. More importantly, there is a form of oxygen called "*ozone*". Ozone protects Earth from ultraviolet light but it is actually poisonous to living things. This is why ozone is found high in the atmosphere so it won't hurt living things. Without ozone, life on Earth would not survive.

The size and spin of the Earth is perfect for life also. If it were any larger, gravity would crush living things. The Earth spins at just the right speed to keep temperatures from being too extreme. The Earth is also tilted at 23.5 degrees. This tilt is what creates our four seasons. Without this tilt, most of the Earth would be terribly hot or cold most of the time.

There are also many forces in nature that hold things together so that life can exist. Gravity is a good example of a force of nature that is perfect for life. If the force of gravity were any stronger, everything on Earth would be squashed. If it were any weaker, everything would float away. The force of gravity could have been much stronger or much weaker but instead, it is perfect for life on Earth.

The truth is, there are at least 30 different forces in nature that all must be perfect for life to exist. If any one of these forces were very much stronger or weaker, life would not be possible.[68] This is called the "*fine tuning*" of the universe. Wherever scientists look, they find these forces in perfect balance.

The "*fine tuning*" of the universe is very tight. The balance between all of the forces of nature is fine tuned to at least "*one part in a hundred million billion billion billion billion billion. That would be a ten followed by fifty-three zeroes*".[69] This is so perfect we can hardly imagine it.[70]

The many forces in nature seem to have one strange thing in common: they are all needed for life to exist. It's almost as if the entire universe was built for just one purpose - life. The idea that all these forces seem to have been built for one purpose is sometimes called "*the anthropic principle*". The word "*anthropic*" comes from the Greek word "*Anthropos*" which means "*man*". In other words, it seems like the whole universe was made just so that humans could live in it.

Some scientists think the universe is just an accident. The problem with this thinking is that accidents don't usually create precision. The chances are out of this world. If we flip a coin 50 times and each time we get "*heads*", everyone would suspect the coin was a fake. We think this way because we know the chances of getting heads 50 times in a row are incredible - about one chance in a million billion. Whenever we see something that is very, very unlikely we suspect it isn't an accident.

The forces of nature in the universe are perfectly balanced so that life can exist on Earth. It would be like flipping a normal coin a trillion times and getting "*heads*" every time. Chance and accident cannot explain a universe so perfectly balanced.[71] In fact, the universe is really an on-going miracle. The forces of nature are constant. Day after day they keep the atoms that make up everything from flying apart. Night after night they keep energy, temperature, radiation and dozens of other forces in perfect balance so that we can continue to live. The chances this universe could be exactly as it is are so huge there isn't enough room in the universe to write out the number.[72]

How to remember this:

> The forces in nature stand on the edge of a knife, it looks like the universe was FIT just for life!

F.I.T.

- **F** = FINE TUNING of the forces of nature is perfect for life.
- **I** = INCREDIBLE odds the universe could be balanced just as it is.
- **T** = TIGHT balance of the forces of nature is tuned to one part in 10^{53} – a number so huge we can barely imagine it.

Question 4

Comets: What is the Oort cloud?

The Oort cloud is an important theory to many Evolutionist astronomers as they study space. This is because Evolutionists are having a hard time explaining short period comets. Already, we know that comets are large chunks of ice and dust that orbit the sun. Comets, however, have an unusual oval shaped orbit which brings them close to the sun and then hurls them far out into space again. Comets also have very regular orbits and have been seen far back in human history.

When a comet gets close to the sun, solar radiation melts some of the ice and dust off of the surface. The melted material streams away from the comet core in a brilliant, fiery tail. The tail can often be seen from Earth, even during daytime.

Some Evolutionists think that microbes in a comet's tail may have landed on Earth and this may be how life got started here. Others think that a comet or a meteor may have hit the Earth millions of years ago causing the extinction of the dinosaurs. But no one has ever seen anything like microbes on a comet. In fact, there has never been any real evidence of life-forms outside of our Earth anywhere, much less on a comet. Comets, however, do have something to say about the past.

Each time a comet orbits the sun and produces a tail, some of it will melt away. For example, in 1986, Halley's comet came as expected, 75 years after its last visit. Its brilliant tail, which was too far from Earth to be seen very well, melted several feet off of the comet's core. Halley's core is about 9 miles wide so it should be there again for us to see in 2061 and again in 2136 and so on for a long time. Eventually, however, it will melt away. If Halley's comet, along with the rest of our solar system, was formed 4.5 billion years ago, why are there any comets left? After 4.5 billion years, even a comet as large as Halley's would have melted away long ago.

A Dutch astronomer named Jan Oort proposed a radical idea to explain comets. He believes there is a giant cloud of "proto-comets" somewhere just outside the solar system. He thinks that every once in a great while the gravity from a passing star might cause of few of these "proto-comets" to sling into our solar system. Once they enter the system, the sun's gravity takes over and they become regular comets.

The Oort cloud solves problems for Evolutionists and, as a result, many astronomers hope to find proof that it exists. So far, there is no evidence for it at all. It has never been seen or even hinted at by any evidence whatsoever. Still, many astronomers continue to search for some kind of Oort cloud.[73]

How to remember this:

Oort tried to show, how the comets could flow, from a cloud that no one has seen. But this cloud is not found, after looking around, less TIME is all it can mean.

T.I.M.E.

- **T** = TALES from comets are the ice melting away as it gets close to the sun.
- **I** = INDIVIDUAL comets should melt away after 4.5 billion years.
- **M** = MORE comets could be floating outside the solar system is what Oort believed.
- **E** = EVIDENCE for this Oort cloud is totally lacking.

Question 5

The Moon: What can the Moon tell us about the history of life?

The four major ideas about how the Moon formed all have serious scientific problems. This presents a problem for Evolutionists because the Moon is essential to life on Earth. Without the Moon, there would be no ocean tides. The tides help stabilize the Earth's tilt (which stabilizes our seasons) and provides a cycle of moving water. This cycle of the tides cleans the shorelines and moves nutrients into the sea. In addition, tides help marine algae grow and marine algae produce oxygen. All of these systems are vital to life on Earth.[74]

The first of the four major ideas about how the Moon formed is called the "fission theory". This idea was first proposed by Charles Darwin's son, George Darwin in the late 1800s. He thought that millions of years ago, the Earth was a spinning mass of molten lava. His idea was that a large glob of molten rock was flung out into space to form the Moon, sort of like mud flying off of a tire. Some scientists thought the Pacific Ocean was the scar left behind. Over the years, however, it became clear that this idea didn't fit the facts. For one thing, the Earth and Moon do not have enough circular motion to fit this idea. Also, if this is what happened, the Moon should be orbiting the Earth at the equator but is, instead, orbiting at 18.5 to 28.5 degrees from the Earth's equator. This is why the Moon appears higher and lower in the sky during the seasons. Also, if the Moon had been made after being thrown off the molten Earth, the rocks on the Moon should be the same as those on Earth – but they are different in many ways. Finally, if the Moon had been flung out into space, gravity would have pulled it back in. If the glob had been too close to the Earth, it would have broken up like the rings of Saturn.

Another idea about where the Moon came from is called the "*capture theory*". This is the idea that a small planet (the Moon) came too close to Earth and was caught by Earth's gravity. There are many problems with this idea that have to do with the speed of the Moon's orbit. There is no known way to slow the Moon down once it came close to the Earth. This is called the "dissipation" problem. If the Moon had been a passing planet but it was moving too fast, it would have just slung away from the Earth. If it had been going too slow, it would have broken up like the rings of Saturn. Even if the conditions had been just perfect, the Moon's orbit would probably have been in an oval shape, like a comet's orbit around the sun.[75] The facts just don't fit the capture theory.

A third popular idea is called the "*condensation*" or "*nebular contraction*" theory. This idea is that the Earth and Moon just happened to form at the same time from very dense clouds of dust. These clouds of dust began to contract into the Earth and the Moon and just happened to be in exactly the right place, with just exactly the right speed and with the perfect balance between temperature and gravity. Even if everything had been just perfect, there is no evidence these kinds of things actually happen. For example, there is a cloud of asteroid particles just beyond the orbit of Mars. This cloud of asteroids shows no sign of coming together into a solid planet. Besides, the amounts of elements we find in Moon rocks is very different from the amounts of elements we find on Earth. These concentrations should be very similar if both the Earth and the Moon came from the same cloud of dust.[76]

The most popular theory today is that a large planet-sized object hit the Earth millions of years ago, throwing billions of tons of debris out into space. This debris eventually condensed into the

Moon. Again, there are serious problems with this idea.[77] Such an impact would have melted the Earth's crust. There is very little chance such an impact would have been exactly the right speed, from exactly the right direction and at the perfect angle required. Even if everything had been perfect, gravity would probably have either pulled the debris back into the Earth or created a ring like we see today around Saturn.

The Moon appears to be in just the right position to regulate the tides on Earth in just the right way to support life. There doesn't seem any natural, accidental way to explain how the Moon formed. All of these things and more seem to say that the Moon was purposely planned to support life on this planet just as the Bible says.

How to remember this:

Moon INFO can ARM us with facts that can aide, our knowing the Earth and the Moon - they were made!

I.N.F.O.

The **I**MPACT Theory, **N**EBULAR theory, **F**ISSION theory, and the **O**RBITAL capture theory do not work.

Knowing this we are **ARM**ed with facts showing the Earth Moon system had to have been made.

A.R.M.

- **A** = **A**NGLE of the moon's orbit doesn't fit these theories.
- **R** = **R**EQUIRED, the Moon is required for life on Earth.
- **M** = **M**INERALS, the mineral elements making up the Moon do not match what we find on Earth.

Question 6

Natural Selection: How Evolution is supposed to work?

The idea that the universe created itself has been around as long as there have been people. In 1859, Charles Darwin tried to explain how life-forms on Earth came about by a natural process in his book *On the Origin of Species*. The idea he came up with is called "*natural selection*".

Natural selection is sometimes called "*survival of the fittest*". It is the idea that most creatures have many offspring that don't survive to become adults because it is only the strongest and best offspring that live. Creatures that are more fit in some way than others survive better. The weaker offspring die out. After millions of years, only the really good varieties of plants and animals will be left. Gradually, tiny changes in each new generation of animals or plants will build up until the creatures are entirely different from what they were originally. This idea is what Darwin described in his book and it is still used today to describe how Evolution is supposed to work. Today most

Survival of the Fittest:
Why fish-nerds went extinct.

scientists would describe Evolution as natural selection leading to new information in a creature's DNA code.

No one doubts that natural selection works in little ways in nature. For instance, you are not exactly the same as your parents. You might be taller or stronger, or you might not.

There is also no rule saying only the best or the strongest offspring will survive.

What if the weak offspring just happen to be the lucky ones? Also, scientists have seen that natural selection is limited. It has boundaries. Animals can have some variety, but the range of variety is limited.

Think about pigeons. There are many varieties of these birds, like the rock pigeon and the passenger pigeon (now extinct), but they are all still pigeons. Darwin bred many generations of pigeons, and no matter how hard he tried, all he ever came up with were more pigeons. Many varieties could come from the original pair, but they were still pigeons! Also, Darwin found out that if he didn't direct their breeding, the pigeons would always turn back into the original common rock pigeon in a few generations. This was true no matter how fancy his first pair happened to be.

No one has ever seen natural selection change one animal into an entirely new animal. No one has ever directed the breeding of a group of animals and found them changing into a completely different kind. Even after thousands of years of selective breeding of animals, dogs are still dogs, cows are still cows and sheep are still sheep.

Evolutionists still think natural selection, or something like it, will produce a new species after millions of years. A species is supposed to be an entirely new creature with new information in its

DNA code, but scientists are still deciding about what exactly is a species in the first place.

Creationists think natural selection can't produce a different animal or plant *kind*. A "*kind*" is an entirely different thing from a species. Scientists will probably argue for years over what is a species, but a *kind* is much easier to understand. There may be many different dog varieties, but there is only one dog-kind. The variations within the dog kind will never produce another *kind* of animal. Any child can tell that a pug and a poodle are the same *kind* even though they have some differences.

It seems much more likely that natural selection is designed to keep an animal population flexible. This means variations will allow the animal kind, like dogs, to survive if the weather or food supply changes. It wouldn't make sense for God to create dogs and other animals that would die out in one generation if the weather changed. What if the weather became suddenly much colder? Probably, only dogs with warm, thick coats would survive. Variation within a kind is probably God's way of helping animals survive in the various climates of the natural world.

How to remember this:

Natural selection is SURVIVAL of the FIT, the limits of variety – that's where it quits.

SURVIVAL of the F.I.T.

- **SURVIVAL** = life-forms have more offspring than can possibly survive.
- **F** = FIRST to survive have the best qualities.
- **I** = INCREASING numbers of good qualities are supposed to build up changing one kind into another
- There is a **T** = TOTAL lack of evidence this has ever happened.

Question 7

Have Scientists Created Life in the Lab? What was the Miller experiment?

All life-forms on Earth use DNA as a molecule code. DNA directs how a life-form will be built and it is very complex. It is like a natural, microscopically small super-computer. It can store as much information as 500,000 type-written pages using small print. It is also so small you could put the DNA for every human being alive today inside a space half the size of a quarter.

The DNA inside your body is constantly being used to put molecules together to keep all your cells strong and healthy. This is what makes you alive. The DNA codes from your mother and father were put together to make your body. This means your body is unique from that of every other human being.

If life started just by accident, that means DNA had to have started by some natural, accidental process too. Most Evolutionists think DNA formed naturally in the Earth's early oceans about 1.5 billion years ago. As we've seen already in this book, the mathematical odds this could happen are about one chance in $10^{40,000}$. These are about the same chances you could run a tornado through a junkyard and see the wind perfectly put together a 747 jetliner.[78] Even worse, scientists know that DNA is used to build proteins but proteins are needed to make DNA – so which came first since you need one to make the other and vice versa? All of this and more tells us that life could not have been built by any accidental, natural process. Still, some scientists continue to believe DNA came about by accident.

Some scientists have tried to make DNA in their laboratories. Dr. Stanley Miller and also Dr. Sidney Fox, among some others, began these kinds of experiments in the 1950s. Some textbooks say that Miller's experiment showed how DNA could have formed. The truth is, no one has even come close to making DNA in a laboratory.

In these experiments, scientists mixed methane, water vapor, ammonia and sometimes sulphur, among other things, in a sealed chamber. They chose these ingredients based on how Earth's early atmosphere might have been made up. Studies on the fumes from volcanoes formed these ideas. No oxygen was included because oxygen breaks molecules apart. It was assumed there was no oxygen in Earth's early atmosphere.

Scientists have toyed with this mixture in various ways over the years. They have used electric sparks in these test chambers to simulate lightning. They have also raised and lowered temperatures and tried other chemical mixtures.

At first, they were pleased to find that a few amino acids were formed in these test chambers. Some scientists have decided that since amino acids are used to make proteins necessary for life, the experiment was on the right track. Some textbooks still report that Miller's experiment created the "*building blocks of life*" in the laboratory. But a few amino acids are a long, long way from DNA. You need hundreds of different proteins in exactly the right sequence to make the proteins necessary for DNA.[79]

These experiments did not show anything close to a DNA chain. Also, no protein *enzymes* were formed, which are needed to build DNA. But, of course, DNA is needed to make these enzymes in the first place. Besides all of this, these experiments were based on ideas about the Earth's first atmosphere that we now know are not true.[80]

The amino acids that did form in Miller's experiment and others like it could never be used to make a living cell because they were the wrong shape. In our bodies, proteins are built from amino acids that are all "*left handed*" in shape. If any of these amino acids are the wrong shape, proteins will not work.[81]

Finally, the Miller experiment and others like it have never made a living cell, nor have they produced any proteins, or cell parts. Scientists have not created life in a laboratory. Every idea about how life could have started by some accidental mixing of chemicals has been shown to be false. Every attempt to mix chemicals in the lab to make life has failed. An honest scientist should see that only a very intelligent designer could have put things together in just the right way to make life work.[82]

How to remember this:

DNA CHEMISTRY makes things alive, from a chemical soup we did not arrive.

D.N.A. CHEMISTRY

- **D** = DNA has never been made in a laboratory.
- **N** = NO **protein enzymes** have ever been made in a laboratory.
- **A** = AMINO **acids** must be left handed for DNA and none of these have ever been made in a lab.
- **CHEMISTRY** alone cannot create life.

Question *8*

The Grand Canyon: Doesn't the Grand Canyon show millions of years in the rock layers?

The Grand Canyon extends across most of Northern Arizona and is one of the largest canyons in the world. It is 250,000 square miles in size and is over a mile deep. It is one of the wonders of the world and beautiful beyond compare. One of the things that stand out about the canyon is the many colored layers of stone in the canyon walls. These layers seem to stretch in unbroken lines for miles.

Many scientists think the canyon shows a record of many different time periods in Earth history. Some people say you can walk from the top of the canyon to the bottom and see fossils showing hundreds of millions of years of Evolution.

Most Evolutionists think that each layer of stone shows a different environment. Each environment supposedly created a layer. When the weather changed, a new layer began to form over top of the old one. Over many millions of years, the layers turned to stone. They think you can see swampy periods, times when the area was a desert, and other times when the area was the bottom of the ocean. Over time, the Colorado River supposedly cut away the canyon.

There are many fossils in these layers. You can also see places where volcanoes poured lava over the top of the canyon. If the Evolutionist idea about how the canyon was made is true, it must have taken many millions of years for the area to form.

Most of the rock layers in the Grand Canyon are made of sedimentary rocks. Sedimentary rocks are normally made when water or wind moves soil and debris. This soil and debris is called "*sediment*" and over time, layers of sediments can harden into stone.

There are a number of serious problems with a slow and gradual idea about how the Grand Canyon formed. In the first place, Evolutionists have a difficult time explaining how the sediment layers were made. For example, there are sedimentary rocks of limestone, sandstone and shale. The problem is that Evolutionists think each of these kinds of sedimentary rocks must have been made in a different environment. This would mean that the entire area - all 250,000 square miles of it - had to have raised and then lowered again and again for each new layer to form. If this is what really happened, the layers would not be so straight.[83]

In fact, the Grand Canyon land area is called the Kaibab plateau. It has been lifted nearly 8,000 feet above sea level sometime in the past. This is amazing because this lifting of the ground did not bend, fold or crack the layers in the Grand Canyon. How this could happen is a great mystery.

"Og discovers the Grand Canyon"

It makes more sense to think all the layers were laid down first and then the Grand Canyon was cut by the river. The problem with this idea is that water just doesn't run uphill!

The source of the Colorado River is high in the Rocky Mountains, high above the Kaibab plateau. As the river flows out of the Rocky Mountains, before it reaches the canyon, it is actually at a lower elevation than the rim of the canyon - by nearly five thousand feet. This means the river would have had to cut uphill 5,000 feet in order to cut the canyon through the Kaibab plateau as we see it today. This presents some serious problems.

If the plateau was raised and then lowered many times, the river would probably have changed directions many millions of years ago. If it had been cutting during each environment change, there would be evidence of erosion between rock layers. But the layers are almost knife edge sharp between each layer. There is no evidence of the million year time gaps that are supposed to have happened between them.[84]

The truth is, the layers in the canyon do not really look like they came from different environments. This is because we can see heavier materials at the bottom of the layers and lighter materials at the top. This is evidence the whole layer was actually made very quickly underwater. When strong underwater currents move gravel, rocks and other debris, the heavier materials settle to the bottom. This is called "*hydrodynamic sorting*".

Strong underwater currents can move many tons of material. When the current slows down, the soil material in the water current can settle to the bottom. If another current moves in at a different speed or carrying a different material, a whole new layer will form over top of the first layer. Scientists have studied this effect in laboratories. We have even seen underwater earthquakes cause huge currents to move millions of tons of mud. These currents can cover gigantic stretches of the ocean bottom with several feet of mud in just a few hours.

The layers in the Grand Canyon look like the layers formed by underwater currents. We can even see ripple marks in the layers. There are places in the canyon where raindrops and animals tracks have been preserved in the stone also. This could happen if a tidal wave laid down a layer and the water pulled back for awhile. After some animals and raindrops marked the ground, another wave brought in a new layer of mud.[85]

Studies have shown that a layer of sediment during a flood or windstorm can be laid down by a current that is faster or slower or coming from a different direction. When this happens, the layers may tilt in many different directions. This is called "*crossbedding*" and we can see examples of "*crossbedding*" in some layers of the Grand Canyon. There are also areas in the Canyon where gigantic boulders are stacked up. Crossbedding, stacking and sorting are things which would happen in a violent flood. The truth is the great, world-wide flood in the days of Noah can explain how the canyon layers were formed better than any modern idea.

Creationists have studied the natural formation of a canyon that happened very suddenly on May 19, 1982. This canyon was made by natural forces on the north fork of the Toutle River in the state of Washington. Creationists think the Grand Canyon was probably made the same way. The Toutle River Canyon was formed as a result of the May 1980 explosion of Mt. St. Helens. After the eruption, tremendous mud flows and ash fallout laid down fine, layered sediments. Nearly 400 feet of ash and mud were laid down in layers in just nine hours. In one of these areas, the ash and mud sediments blocked the flow of a small creek. This created a dam several hundred feet deep. Within a short time, water built up a large lake behind the dam. Eventually, on May 19, 1982, the lake overflowed the dam with tremendous force. The water pouring over the top of the dam cut a huge canyon 140 feet deep. Afterwards, the small stream was left in the bottom of the canyon and the layers of ash and mud became as hard as stone within just five years.

Any modern geologist, not knowing the facts, might think the fine ash layers in this new canyon came from many different eruptions of Mt. St. Helens instead of just one eruption. A scientist who didn't know what happened might even decide the mud layers might have come from an ancient lake bottom. If he didn't know better, he might then say the creek carved the canyon over many years of slow and gradual erosion.[86]

There is strong evidence the layers in the Grand Canyon were laid down quickly. Since the layers in the Canyon are so straight, it makes sense to think they were cut away quickly before the layers had time to completely harden into stone. During Noah's Flood, there were probably many giant tidal waves. There were also earthquakes, volcanic eruptions and huge mudflows. It is possible a lake as big as the Great Lakes today formed behind the Kaibab Plateau. The Grand Canyon may have been cut away quickly when this enormous lake poured over the Kaibab Plateau Dam. This explains the underwater sorting we see in the Grand Canyon sediments. It also explains why the sediment layers show no evidence of time between the layers.[87]

The Grand Canyon does not show many different environments over many millions of years. Instead, it shows what happened when God destroyed the world with a mighty flood.

How to remember this:

Flood waters ran over and made quite a bite, The Grand Canyon resulted and it's quite a SITE.

S.I.T.E.

- **S** = STRAIGHT layers of rock show no signs of being raised or lowered in the past.
- **I** = IRREGULAR sized sediment is found at the bottom of layers made underwater.
- **T** = TOP. The top of the canyon is above the flow of the river so the river would have had to run uphill to cut the canyon.
- **E** = EXAMPLE. An example of a canyon that formed when a lake overran a natural dam can be studied at Mt. St. Helens.

Question 9

Transitional forms: Why are transitional forms so important?

The Theory of Evolution is the idea that all life, from bacteria to dinosaurs, developed very slowly by tiny changes in animals and plants, generation after generation. Evolutionists think these tiny changes, also called "*mutations*", slowly transformed one type of animal or plant into another. All of this slow and gradual change should have taken many millions of years. Hundreds of millions of life-forms would have lived and died between one type of animal or plant and the next. If Evolution is true, scientists should be able to see millions of life-forms that are in-between one life-form kind and another. In-between forms are also called "transitional forms".

After more than a hundred years of study, scientists have only found a small handful of fossil creatures that might be transitional forms. The truth is, there are no *clear* examples of transitional forms at all. The very few fossils Evolutionists talk about are not very clear. In fact, some of these possible transitional forms are probably just different species and some have even turned out to be fakes.

It is true that there is great variety within living creatures. This variety comes from the mixing of the DNA codes of the parents. If one parent is tall and the other is short, the offspring may get one code for tall and one for short. Mixing and matching like this creates variety but the basic information in the code stays the same. This is why you can have tall humans and short humans, but you don't see humans born with wings. Having wings would be completely new information in the DNA code.

Every once in a while mutations happen in the DNA code. Evolutionists think these tiny mistakes might build a totally new code. Normally, mutations are very harmful. For example, sickle cell anemia is a mutation in the DNA code of humans. Most people with this mutation have serious medical problems and usually don't live very long. Mutations like sickle cell anemia are normally weeded out over time because the mutation doesn't help the offspring survive.

Evolutionists think there must have been "*good*" mutations in the past that helped life-forms survive better. When the environment changed, they think life-forms with the new DNA information survived better than life-forms with the old information. After millions of years, many of these mutations supposedly caused one kind of life-form to transform into something completely different.

Now, it is obvious that variety in living things allows animal and plant kinds to survive as the environment changes. This is called "*adaptation*" but adaptation is not new information in the DNA code. Life-forms do need some ability to adapt or else they would quickly be destroyed if the weather changed. The real question is whether or not variety can lead to new information in the DNA code. In other words, is *variety* evidence that life-forms are changing from one kind into another or does *variety* have a limit? If there is a limit, variety could be seen as evidence of good design instead of Evolutionary change. There are only two ways to find out.

First, do we see Evolution happening today or do we see that life-forms have limits in their variety? Second, if Evolution is true, the fossil record of the past should show millions of examples of transitional forms. On the other hand, if Evolution is not true, the fossil record should show that life-forms in the past had limits to their variety also.

In order to understand this clearly, think about whales. Whales are examples of the largest creatures ever to live on earth. Because they are mammals that live in the ocean, they breathe air instead of using gills. Unlike some fish, whales are born alive and feed their young with milk just as other mammals do. They are also warm-blooded.

Whales are interesting also because they have skin that is very different from other mammals. When other mammals, such as human beings, spend too much time in the ocean, they begin to dehydrate. Every child knows that spending too much time in the water makes your skin wrinkle. These wrinkles come from dehydration in the salt water. Whales, however, do not experience this problem because their skin is specially designed for life in salt water.

Whales are also insulated from the cold. As warm-blooded animals, they require large amounts of food in order to keep up their body temperature. Some whales have teeth for catching food and others have giant straining nets of baleen in their mouths used to collect plankton and shrimp. They also have layers of fat to help keep warm.

As with bats, whales also have a special echo-listening or "*sonar*" system allowing them to find their way under water. They also have a different kind of inner ear from other mammals and hollow places in the skull so they can dive down into the great depths.

Even more amazing, whales have special blood that is designed to pick up more oxygen than regular mammal blood. They also have very rich mother's milk. Whale mothers have special mammary gland muscles so that baby whales can feed quickly underwater and not drown. Even more amazing, some whales have "*heat exchanging systems*" in their blood vessels. These systems keep thin areas of the body that are not covered by fat from freezing in very cold water.[88]

Evolutionists cannot imagine any step-by-step way these amazing features could have come from a land creature by tiny accidental changes over millions of years. It takes a great deal of belief and imagination to think that a four-legged mammal such as a cow or a wolf somehow had offspring with accidental tiny changes that allowed it to crawl into the ocean and survive.[89]

Over the years, Evolutionists have identified several fossil creatures they think could be the land-dwelling animal that became a whale. They think these creatures may have been the "*first whales*" because their teeth look something like primitive whale teeth. Still, just because a land-dwelling animal had teeth that look something like primitive whale teeth doesn't prove they gradually turned into whales over many millions of years.

If Evolution is true, the "*first whale*" must have been born with specialized skin to avoid dehydration right from the start. The first whale would also need specialized mouth parts to feed itself and a way to breathe while waiting millions of years for its nose to move to the top of the head. How did it keep from drowning while waiting for a genetic accident to develop the muscles needed to cover its blow hole while breathing and diving?

If the Evolution story for the whale is true, whatever land animal began the process must have had hind legs that gradually became useless and disappeared. Even if we could imagine such a story, how could the "*first whale*" possibly survive on land while dragging around useless legs? Why would genetic accidents cause legs to disappear or create an enormous, muscled tail with flukes while the creature was still a land animal? Put the other way, if all of this change happened while the "*first whale*" was already in the water, we should see clear evidence of the animal's nostrils gradually moving up the skull from the nose to the top of the head where we find the breathing hole in modern whales. How could this animal survive in the water and find food without the special sonar found in living whales? How did this animal survive while gradually developing the specialized baleen netting structures the great baleen whales use for straining plankton and shrimp?

Obviously, after millions of years, scientists should be able to find many transitional forms, or in-between creatures, between a land animal ancestor and the whale. There should be plenty of examples of land-dwelling creatures with tails and flukes as they wandered into the water.

The only "*candidates*" for whale Evolution are either obvious land-dwelling animals or interesting creatures that look like the modern day walrus. One animal some Evolutionists think was one of the "*first whales*" is called *Mesonyx*. This animal was pulled out of the fossil record because its teeth appear similar to some kinds of whale teeth. This is difficult to believe since *Mesonyx* was a land-dwelling animal similar to a wolf. It also had hooves.

In fact, many of the supposed ancestors for whales are only known from fossil teeth and scientists have absolutely no idea what the rest of the animal looked like.[90] Evolutionists really want to find a transitional form between land-dwelling animals and whales but there are no clear step-by-step transitional forms for whales. Occasionally, Evolutionists just decide that something like *Mesonyx* must be a whale ancestor simply because it had peg-like teeth, but just calling something a transitional form and showing it really is one are two different things.[91]

The truth is, the only possible transitional forms Evolutionists bring up are not very convincing. There are some interesting mosaic creatures like the platypus and the walrus but these creatures are complete within their own kind and are not transitional forms. If Creation is what really happened in the past, this is what we would expect to find in the fossil record.[92]

The fossil record is very rich in examples of animals and plants from the past but all of the major kinds of life-forms are found complete within their kind. We should expect some gaps in the record since fossilization is supposed to be a random process but these gaps should be at random. Instead, we find a consistent pattern of huge gaps between every major life form. It appears, at least from the fossil record, that the major life-form kinds have a great variety within themselves but have never changed into completely different life-forms over time.[93]

How to remember this:

> **TRANSITIONAL** forms can make the case, they put evolution **BETWEEN** a **ROCK** and a **HARD PLACE**.
>
> TRANSITIONAL forms, also called in-BETWEEN forms, are the creatures that should exist step-by-step between one kind and another kind. In-between fossils are not found in the ROCKs and this puts Evolution in a HARD PLACE.

Question 10

The Cambrian: What was the Cambrian Explosion?

The idea of Evolution is that the first kind of life-form changed very slowly over millions of years into every other kind of life-form. This idea is sometimes called "*descent with modification*". If this is really what happened in history, we should see a long chain of change. This change would have started with single celled life-forms like algae. The next step would have been many-celled animals without backbones, then fish, amphibians, reptiles, mammals and ultimately -people.

The chances that the first life-form could have been put together by accident are very small. The odds would be about the same that a person blindfolded could toss a dart from Pluto and hit a target on Earth the size of a proton on the first try. Modern attempts to create a basic life form from chemicals in a laboratory have all failed as well. It appears impossible for life to have come about from a chemical soup here on Earth.[94]

Some scientists insist that we just have to keep studying until we eventually figure out how life started from a chemical soup even though the laws of chemistry say it cannot happen. Even if we skip over this question and just believe life somehow got started by accident, there are still some very serious problems with the Evolutionist story.

For example, according to this story, millions of years after the first single-celled life-form was born, these single-celled creatures suddenly changed. They became a wide range of soft-bodied, invertebrate animals (animals without backbones) such as sponges, jellyfish, sea cucumbers and sea lilies. This sudden change supposedly happened 600 million years ago. Evolutionists call this the Cambrian time period. The Cambrian change from the first kind of single-celled life-forms into the Cambrian life-forms supposedly took 80 million years of Evolution. But just like every other part of the Evolution story, there are many problems with believing it really happened.

In the first place, Evolutionists think the first life-forms on Earth formed between 1.5 and 3.5 billion years ago. If the Cambrian change happened 600 million years ago, there is at least a one billion-year gap, or longer, between the first life-form and the first great change. There is, however, almost nothing in the fossil record before the Cambrian. In fact, the change during the Cambrian time period is so sudden that many scientists call it the "*Cambrian explosion*".

"The REAL Cambrian Explosion"

Rocks that scientists believe are from the Cambrian period are filled with some very complex undersea creatures. There are worms, trilobites, snails and other amazing animals. Each of these creatures is found without any trace of ancestors in pre-Cambrian rocks. If the pre-Cambrian life-forms were around for at least one billion years, there should be plenty of evidence of the ancestors of the Cambrian life-forms – but there isn't.[95]

Many-celled life-forms such as jelly fish and sea cucumbers are much more complicated than single-celled life-forms. A many-celled creature has cells that work together to do things such as digestion and movement. These creatures also use amazing chemical and electrical impulses to get their cells to work together. There should be billions of fossil in-between forms between "*simple*" life forms and the animals without backbones we see in the Cambrian. There is no evidence these in-between forms ever existed.

The Cambrian explosion of life is so sudden and so vast, it completely baffles Evolutionists. If Evolution is what happened in the past, Evolutionists need to explain how this happened.[96]

Some scientists try to explain the lack of in-between forms by saying the life-forms from this time period probably didn't fossilize very well. This isn't a very good explanation because we have thousands of examples of soft bodied creatures like jelly fish that have fossilized very well. The fact these in-between forms do not exist is only proof that they do not exist - not proof they didn't fossilize well or evolved too rapidly to leave a trace.[97]

Creationists believe God created everything, from single-celled life-forms to dinosaurs, using just His own power. Creationist scientists who believe the Bible explain that there was a world-wide flood during the days of Noah. This flood caused huge tidal waves to scour the continents in wave after wave of destruction. This would mean the coastal areas and shallow continental shelves where many invertebrates lived were probably covered over first. When water and mud is stirred up, certain shapes and weights of objects will settle to the bottom. This is called "*hydrodynamic sorting*" and this process can explain why invertebrate fossils formed in large numbers in the lowest layers of mud from the flood. Bible-believing Creationists think that most of the fossils we find today came from the Great Flood. When the evidence is examined it appears that jellyfish have always been jellyfish and clams have always been clams.[98]

How to remember this:

The Cambrian Explosion tells us a ton, about in-between fossils of which there are none: One SINGLE jellyfish BLOWING APART, would make many MULTIPLE JELLYFISH P-AR-T=S.

SINGLE celled life-forms supposedly changed so suddenly it was like BLOWING APART into many MULTI-celled life-forms such as JELLYFISH.

P-AR-T=S

PROBLEMS - ARE –TRANSITIONAL forms = SERIOUSLY lacking.

Question 11

Living Fossils: What is a Coelacanth and why is it important?

The theory of Evolution teaches that all life-forms today came into existence over millions of years. This supposedly happened through a long series of tiny changes between one generation and the next. These tiny changes caused the first single-celled form of life to transform into a many-celled life-form such as a jellyfish. More accidental changes in their DNA caused the many-celled life-forms to become fish, then amphibians, then reptiles, then mammals and ultimately, people.

Evolutionists need to find the in-between forms, also called transitional forms, for each of these steps to prove their theory is what really happened. One of the most important life-form changes in this story would be the change from a fish into an amphibian. Fish, of course, are perfectly designed for life in the water. Amphibians are very specialized animals that have both gills and lungs and all the extra features needed for moving about on land. These two animals are completely different creatures except that they both can live in the water.

Evolutionists searched the fossil record carefully to find a fish that could be the ancestor for amphibians. They chose the *Crossopterygian (Cross – op –ter – ee – gee – en)* or bony fish. The Coelacanth (See – low – kenth) is a crossopterygian fossil fish that scientists thought was the ancestor of all land-living animals today.

Coelacanth fossils had some features that looked a little like the amphibians. These features include arch-like bones in the backbone and a few features that looked a little like an amphibian in the skull. They also had "*lobe*" fins Evolutionists thought could have changed into legs. Over millions of years, these fish were supposed to have become the first amphibians by slow DNA mistakes in the fish population. One Evolutionist even said we ought to get over it and start calling humans "*specialized fish*" since we all came from these crossopterygian fish.[99]

This idea was thrown a curve ball in 1938 when a living coelacanth was caught by a fisherman off the coast of Africa. Evolutionists thought this kind of fish had changed into all land animals and had gone extinct 70 – 80 million years ago. Since 1938, coelacanths have been found around Indonesia as well and scientists have done detailed studies on these amazing fish. Evolutionists still believe that all land-living animals came from crossopterygian fish like the coelacanth, but there are many problems with this idea. Now that we have living coelacanths to study, these problems are only getting larger.

In the first place, the coelacanth and other crossopterygian fish are very different from amphibians, as are all fish. It would have taken a true miracle to change a fish into an amphibian.

For example, fish have small, loose pelvic bones that are fixed in muscle. All amphibians, on the other hand, have large pelvic bones that are solidly attached to the backbone. This is so because amphibians carry the weight of their bodies solidly on their legs. On the other hand, fish only have pelvic bones for muscle and organ attachment purposes.

The internal structure of extinct fish and amphibians is not known but we can study living coelacanths. Coelacanths have normal internal organs for a fish that are not in any way in-between a fish and an amphibian. The fins and muscles of the coelacanth are also exactly what we would expect to find for a fish. Coelacanths use their fins to swim just like any other fish.[100]

Even if we could believe that a fish like the coelacanth was the ancestor of all land-living animals, it is impossible to figure out how tiny DNA changes could allow a fish to develop lungs. A fish out of water will die, so obviously, fish had to have developed working lungs while still in the water. Even if we thought this could actually happen, and there is absolutely no evidence that it can, why would they do this? What purpose would lungs serve except to put an animal on land? Even if the world was rapidly drying out 100 million years ago, how could fish "*decide*" to have just the right mistakes in their DNA code to create lungs? No one has ever seen anything like this happen and there is no evidence in the fossil record that this happened either.

What scientists have seen is that there is a living, swimming population of crossopterygian fish existing unchanged in the deep waters of the Indian Ocean. The coelacanths living today just appear to be big, ugly fish. There are no in-between, transitional fossil forms between fish and amphibians. There is no evidence in living coelacanths of any physical part transforming into an amphibian. The evidence in the fossil record shows that fish have always been fish and amphibians have always been amphibians.

But the coelacanth is not the only example of a "*living fossil*". Graptolites, small water-dwelling animals that probably live in colonies, were supposed be extinct for 300 million years until found alive in the South Pacific Ocean. The "*dawn redwood tree*" was supposed to be extinct for 20 million years until they were found alive. Tuatara lizards were found alive in New Zealand but it was supposed to have been extinct since the time of the dinosaurs.[101]

Evolutionists, however, are not quick to give up hope. Even after careful study of the living coelacanth, many Evolutionists still believe crossopterygian fish transformed into all land animals. Today many Evolutionists have turned to another fish from the order *Rhipidistia.* But just like the Coelacanth, this order of fish also has loose pelvic bones and no hint of a developing shoulder.[102] They just call this a mystery of Evolution.[103] For Creationists, the Coelacanth and other fish like it is no mystery. It is, in fact, exactly what they predict scientists should find in the natural world.

How to remember this:

> Coelacanth F.I-SH still live in the sea, it's foolish to think they became you and me.

F.I.-SH

> FOSSIL fish once IMAGINED to be the ancestors of all land animals are STILL HERE.

Question 12

Evolution today: What is the English Peppered Moth?

Since Evolution is the idea that life-forms completely change into different life-forms over time, we should be seeing Evolution happening all around us. If this is true, however, why don't scientists see monkeys evolving into higher primates or reptiles changing into birds today?

Usually, scientists answer this question by saying that Evolution is such a slow process we can't really see it happen. It is, of course, impossible to study an animal population for several million years as a research project. Even Evolutionists who believe in rapid Evolution, sometimes called "*punctuated equilibrium*", explain things this way because "*rapid*" Evolution could still take thousands or tens of thousands of years. "*Rapid*" Evolution would still be far beyond the lifetime of a scientist.[104]

Still, there are a few examples Evolutionists think are Evolution in action. For example, wisdom teeth in humans appear to be dying out. Some people have a full set of wisdom teeth, others get a partial set and others get none at all. Some scientists think humans are becoming a life-form that does not have wisdom teeth.

Wisdom teeth in humans don't really show Evolution at all. For one thing, humans are still human whether they have wisdom teeth or not. If this were really Evolution happening, we would expect to see some kind of new information in human DNA. Instead, humans without wisdom teeth are *missing* a feature, not *gaining* some kind of new and better DNA information. Wisdom teeth are actually just an example of variation within a kind, similar to skin color or height. In fact, there is evidence that wisdom teeth "*crowding*" is the result of poor diet among other non-DNA reasons.

In most public school biology textbooks today, Evolutionists talk about the English peppered moth (*Biston betularia*) as an example of Evolution happening today. The peppered moth was first commonly known as a whitish, light-colored moth fluttering about the grey English skies. In 1848, a black form of the moth was seen and given a new species name, *Biston betularia carbonaria*. Except for the dark color, this moth is identical to the lighter colored variety.

The exciting event for Evolutionists was the shift in moth population in the next few decades after 1848. As the Industrial Revolution ate up the English countryside, smog and smoke from factories began to have an effect on the local environment. Lichen growing on trees began to die, smoke obscured the sun and dark soot settled on everything and everybody. Suddenly, having a light color made the light-colored *Biston betularia* an easy target for hungry birds. Within a few short years, the rare *carbonaria* ended up becoming 90% of the population. Evolutionists thought this change in population was an excellent example of an Evolutionary change from a white moth to a black moth.

There are some serious problems with this idea. Like wisdom teeth in humans, the DNA code for dark coloring was in the moth population from the very start. The black moth has never been shown to be a mutant of any kind.[105]

In other words, the black moths are just an expression of genes on their DNA code that the moths already had. The dark form of the moth was not an example of a DNA mistake producing a new, better individual. The dark form of the moth is just as much an English peppered moth as any other English peppered moth.

Secondly, the light colored variety of the peppered moth still exists. In fact, in recent years, as the effects of pollution are cleaned up, the light colored variety is steadily increasing in numbers. Also, the dark form of the moth is still a moth. It did not transform into an entirely new creature. The peppered moths are still peppered moths whether they are black or white.

The English peppered moth actually shows us that a wide variety of colors and sizes in an animal kind is a good way of helping the animal survive when the environment changes. Creationists think it makes sense that God would design animals and plants this way.[106]

Over the last few years the story of the English peppered moth has started to fall out of favor. This is because the main scientists who studied the moths, Dr. E. B. Ford and Dr. H. B. D. Kettlewell, were later accused of faking some of the evidence. Even if they didn't fake evidence, the science itself is flawed. It was never really proven if the moth population changed because of birds eating more of the light-colored variety or for some other reason.[107]

Most Creation scientists never had a problem with the story in the first place. This is because there was never any new, improved information added to the moth's DNA code and the moths were still moths at the end of the day. In fact, most Creationists see this story as evidence of God's master plan to help a species survive changes in the environment.[108]

How to remember this:

> English peppered MO-T-H-S are still on the scene, no new information was found in their genes.

MO-T-H-S

- **MO** = MOSTLY white moths shifted to mostly dark.
- **T-H** = TOTALLY new DNA information did not HAPPEN.
- **S** = STABLE DNA means the English peppered moth is still an English peppered moth.

Question 13

Dinosaurs: What happened to the dinosaurs?

The most natural reading of the Bible tells us that God created everything, including the dinosaurs, in six days. There are some people who think there is a gap between the first two verses of the Bible's Creation story that was many millions of years long. People who believe in this "*gap theory*" think that dinosaurs and other extinct creatures lived during this gap. There is, however, no real evidence of any kind of gap of millions of years in the book of Genesis.

Other people think the six days of the Creation story were actually periods of millions of years. This is called the "*day-age theory*" but again, there is no evidence in the Bible that the editor of Genesis meant anything other than normal days when he recorded the story. Neither the gap theory nor the day-age theory is well supported by scriptures in the Bible.

There are many honest, well-meaning Christians who believe in the gap theory and the day-age theory. We do need to remember that how we think about Genesis is not what makes a person a Christian. Still, if we doubt the clear ideas written in the first few verses of the Bible, it may cause us to question other important teachings in the Bible too. There is no scientific reason we should try and fit in gaps or days really being millions of years into the Creation story.

What all of this means is that if we accept the Bible's story just as it is written, then dinosaurs must have lived at the same time as human beings and been part of the story of Noah and the Flood. It means that most of the fossils we find today came from the Flood. It also means dinosaurs were on the ark and survived after the Flood.

Evolutionists, of course, think this is a silly idea because they believe the Earth is 4.5 billion years old. They also think dinosaurs and humans lived 65 million years apart. Most Evolutionists think the Noah's ark story is a myth; they can't imagine Noah dealing with something as terrible as *Tyrannosaurs rex* on board the ark. How could Noah handle a brachiosaurus dinosaur that weighed hundreds of tons? And what about the many animals that had specialized diets? How could they survive on the ark? Even if they were to try and look at the story seriously, most Evolutionists point out that it would be impossible.

To begin answering these questions, it is important to keep in mind that only about five percent of the fossils we have found so far are land-dwelling animals and far less than one percent of these are dinosaur fossils. There is much we do not know about dinosaurs. If they were like modern reptiles, it is possible that dinosaurs never stopped growing. Of course, we have found dinosaur eggs. The eggs show us that even the largest dinosaurs started out not much bigger than a large lizard.

There is nothing in the Genesis story telling us that Noah took only *adult* animals either. It is likely that immature dinosaurs were probably easier to handle than the full grown model and also a great deal smaller. Zoo keepers today can handle immature modern crocodiles, lizards and snakes with care.

Most importantly, what do scientists really know about dinosaur behavior? Modern movies show these animals as ferocious monsters, giant eating machines that would have wiped out any human population easily. But what do we really know? After all, for nearly one-hundred years, every scientist *knew* that dinosaurs were slow-moving, cold-blooded swamp creatures with brains so small they were likely as dumb as rocks. It is only very recently that dinosaurs have been cast as fast-moving and smart.[109]

Most modern ideas about dinosaur behavior are based on the remains of bones and footprints. It is difficult, almost impossible, to tell from bones how these creatures actually behaved.

For example, most people think that the famous *Tyrannosaurus rex* was a horrible killing monster who ruled the world with his astounding speed and relentless hunger. "*T*" rex did have six-foot jaws and six inch teeth. There may be, however, another way to explain the bones and teeth of "*T*" rex. In fact, this new explanation is gaining popularity even with Evolutionists like paleontologist Dr. Jack Horner.

The front limbs of "*T*" rex were practically useless. In fact, rex couldn't even reach his mouth to pick his own teeth. They would have had little use for a predator. Some research shows the teeth in rex's jaws may have been set too shallow in his jaw bones to have ripped hard flesh and bone. If *Tyrannosaurus rex* had taken a mouthful of live duckbill dinosaur, he might have had most of his teeth torn out of his gums. The teeth themselves also show very little wear and tear - even after "*65 million years*".

It is possible that *Tyrannosaurus rex* was a scavenger, feeding on the dead remains of fallen animals much like hyenas do today. Hyenas are very sneaky, easily-startled scavengers and "*T*" rex may have behaved in a similar way.

There are certainly other possibilities but scientists have less than 20 incomplete *Tyrannosaurus rex* fossil skeletons to work from anyway. It is also interesting to note the size of the average dinosaur brain case. They had brains about the size of a walnut, which, in comparison with their size, was very small. There is no way to know the true intelligence of these animals. It may very well be that many predatory dinosaurs were much like the hyena is today. Also, many plant-eating dinosaurs may have been herding creatures like elephants and bison are today. Humans have co-existed with lions, bears, mountain lions, alligators and herds of bison for thousands of years. Most animals in the wild today have an instinctive fear of humans. There is no way to know how dinosaurs may have reacted to human beings.

Noah may have understood dinosaurs from his own experiences with them in his world. It may also be, like many animals in the modern world, that dinosaurs were able to hibernate. If so, they may have been much easier to handle as young animals on board the ark. Scientists do not know for certain if they were cold-blooded reptiles, but if they were, cooler climates make reptiles significantly slower also.

Dinosaurs may have acted very differently than modern scientists think that they did. Noah could very well have been able to handle a young *Tyrannasaurus rex* with the same normal caution he would have used handling his lions and tigers and bears.

Why didn't dinosaurs survive the Genesis Flood?

If dinosaurs lived before Noah's Flood along with humans, what happened to them? If they were on the ark, where are they now?

The fossils of plants and animals are all that remain today of the world that Noah knew before the Great Flood. They tell us that prior to the time of the Flood, much of the world was a tropical or semi-tropical environment like Hawaii is today. The story in Genesis hints at a massive invisible water vapor canopy high in the atmosphere which may have helped produce a greenhouse effect and higher air pressure. The description in Genesis seems to say that much of Earth's water supply was underground, allowing large amounts of water to spring up under pressure from numerous springs. Although most of the Earth was tropical or semi-tropical, it is not unlikely that every environment existing today also existed in Noah's time.

The Genesis story tells us about terrible earthquakes at the beginning of the Flood. These earthquakes probably released great underground reservoirs of ground water. Genesis calls these reservoirs "*the fountains of the great deep*". This breakup of the Earth caused gigantic eruptions of volcanoes such as Earth had never seen before or since, sending trillions of tons of ash into the air, blocking out the sun. We can see evidence of this in places such as the Columbia River Lava Plateau where there are 200,000 square miles of lava flows; some of them are miles deep. Nothing has been seen like this in modern times.

At the same time the Earth was breaking up, water vapor in the atmosphere probably collapsed because of the ash in the air. Earthquakes and volcanoes caused gigantic tidal waves hundreds of feet high. These waves would have surged around the Earth, tearing billions of tons of mud and debris off of the continents and laying down thousands of feet of muddy sediments. The resulting Flood lasted 371 days.

After the Flood, the Earth was very different. The Flood had warmed up the oceans, causing many years without summers, heavy snowfalls and huge shifts in temperatures so that snow would alternately pack down and then melt. Several thousand feet of snow fell in the polar regions and the Earth experienced an ice age.

Many animals that came off the ark found they could only survive well in certain areas of the new world where there was enough food or where it was too cold. Right after the Flood, there was intense competition for food. Many animals were forced into extreme environments like the arctic or the desert to survive. Some animals were designed with an ability to survive in extreme environments like deserts and others were not. Dinosaurs and many other extinct varieties of large, land-dwelling creatures that required huge amounts of food to live, were not able to survive very long in the new world.

Some dinosaurs probably did survive for a long time, especially marine reptiles. Most of these animals were probably very rare on the Earth after the Flood and may have lived for long periods, continuing to grow larger like modern turtles do today. It is possible the world's many dragon legends refer to sightings of these amazing animals. These legends often describe dragons as snake-like or lizard-like with bony, armor plating and there are dragon legends from every continent on Earth with these common threads.[110]

Alexander the Great was involved in one of these dragon legends. In his story, he is said to have gone to see a dragon kept in a large cave in India. The monster's fierce appearance and terrible roar scared his group away.

There is another legend about a farmer in Italy who killed a small dragon about the size of a sheep, which had startled him along a roadside. Still another story tells of three knights who set out to kill a famous dragon in a remote part of France in about the 8th century A.D.[111] There is the thunderbird story from North America, the many dragon tales in China and Japan, and the great sea monster stories told in many a sailor's tavern.

These stories, of course, were probably embellished over time, adding magical powers to the dragons. It is hard to say which part of a tale is based on fact and which is based on fearful imagination but most dragon descriptions fit what scientists find in the fossil record. Many dinosaurs had bony plate armor, were scary and large, had terrible teeth and claws. Some had wings or resembled snakes too.

Even today, there are stories about dragons and sea monsters which might have some basis in fact. Keep in mind that the vast majority of the Earth is covered by water that is so deep scientists cannot monitor or explore it very well. No one really knows what kinds of animals lurk in the great depths.

Of course, stories do not prove anything. They are just interesting tales which may or may not turn up something unknown to modern science. Some of these strange stories, however, have uncovered unknown creatures and only faith in Evolution keeps people from fully looking into these leads.

As an example, native people in Rhodesia, Africa told legendary tales about giant apes for many years. Western scientists ignored the stories as native superstition until the early 20th century when the rare highland gorilla - a true great ape of enormous size - was discovered by Western scientists.

Another creature people once thought was extinct since the time of the dinosaurs is the Coelacanth. This large fish was discovered alive in the Indian Ocean by a fisherman in 1938.

The Komodo dragon is a huge meat-eating lizard that can grow over 12 feet long and still lives on a small Indonesian island. The Komodo dragons were only known by legend for many years. Today, the Komodo dragon is nearly 99% extinct but many zoos are trying to breed them.

There are also two references to dinosaur-like creatures in the Bible. Many Bible scholars think the oldest book in the Bible is the Book of Job. Job describes two terrible animals. Obviously, these animals were very rare in Job's day, which was probably almost 3,000 B.C., but they were certainly not unfamiliar creatures to Job.

"*Look now at the Behemoth, which I made along with you; He eats grass like an ox. See now, his strength is in his hips, and his power is in his stomach muscles. He moves his tail like a cedar; the sinews of his thighs are tightly knit. His bones are like beams of bronze, his ribs like bars of iron"* (Job 40:15-18).

"*Can you draw out Leviathan with a hook...Who can open the doors of his face, with his terrible teeth all around? His rows of scales are his pride...They are joined one to another, they stick together and cannot be parted. His sneezing flashes forth light, and his eyes are like the eyelids of the morning. Out of his mouth go burning lights; Sparks of fire shoot out*" (Job 41:1, 14-19).

Some people think the Behemoth is just an elephant or a hippopotamus but neither of these creatures have tails like cedar trees. A cedar can grow well over 100 feet tall and can be as thick as 15 feet wide at the base.

Other people believe Leviathan is just an alligator or a crocodile but the description includes "*sparks of fire*". There is an insect today, called the bombardier beetle, that has a special defense against its enemies. The beetle has two glands in its abdomen than can mix a chemical compound that will explode with a flame, steam and 212 degrees of heat right in the face of its enemies.[112] Perhaps the Leviathan had something like the bombardier beetle's defense.

It is interesting to see that there are many dinosaurs, such as *Corithosaurus*, that had strange, hollow, bony crests on their heads. It is possible these crests had a bombardier beetle-like chemical defense system. This could be what Job was talking about, but he is certainly not talking about an alligator or a crocodile.

There is no way to know how long dinosaurs survived after the Flood. Many animals simply did not survive long in the new world. Even today there are many animals going extinct every year as the environment changes. Myths, legends and living fossils don't prove the story of Noah, but these stories do remind us that there is much we do not yet know.

How to remember this:

Dinosaurs on board! Go get a sword.

Dinosaurs on board

DINOSAURS were ON BOARD Noah's ark, probably as immature animals.

Go get a sword

GOing out of the ark, many dinosaurs died in the new climate but some may have survived for many years.

Sightings of these rare animals are probably the basis of many dragon legends in which they are often killed by a knight's SWORD.

Question 14

Birds: Is Archaeopteryx a transitional form between reptiles and birds?

If the process of Evolution is a fact of history, there should be many obvious in-between forms, also called transitional forms, between one distinct type of animal or plant and another. For example, many Evolutionists today believe that small dinosaurs were the ancestors of modern birds. If this is true there should be some transitional forms between these two groups of animals. Our modern culture has so accepted this theory as a fact that some blockbuster movies today even add feathers to computer generated pictures of dinosaurs. The artwork in many print books and magazines add feathers to some dinosaurs as well even though the actual fossils of these animals do not have feathers.

There is no clear evidence that dinosaurs evolved from birds. We have found dinosaurs with fiber-like fuzz on the skin but this fuzz is found on the wrong kinds of dinosaurs (lizard hipped, non-theropod dinosaurs). Also, this "*dino-fuzz*" has none of the features of feathers.[113]

The belief that dinosaurs turned into birds over millions of years has become so strong that scientists can be easily fooled. In 1999, National Geographic magazine produced an article and a model of Archaeoraptor, a "*feathered dinosaur*" found in China. Finally, scientists had a fossil showing the transition between dinosaurs and birds. Eventually, however, it was discovered that Archaeoraptor was a fraud. The fossil material was found to be a dinosaur tail fossil glued to a bird fossil.[114]

The dinosaur-to-bird change was supposed to have started about 150 million years ago. This change supposedly began with a group of very small theropod dinosaurs that lived on the ground and probably ate insects. The story is that by accidental mistakes in their DNA code, some of these small dinosaurs were able to jump higher than their neighbors so they survived better. Somehow these high-jumping dinosaurs began to develop longer and longer scales. Some scientists think these longer and longer scales had something to do with body temperature. Other scientists think that longer scales might have helped the therapods catch insects better. Evolutionists believe that over time the scales became feathers and the jumping reptiles learned to fly.

There are a number of problems with this story. In the first place, birds are very specialized animals. One of the most specialized features of birds is their feathers. There is a vast difference between the way a scale is made and the way a bird feather is made. There is no evidence in the fossils showing a change from a scale to a feather. There is no way to explain why scales would mutate into feathers when the DNA code for scales only makes scales, not feathers. Most scales on reptiles are, essentially, folds in the skin but feathers develop from tiny follicles in the skin. In fact, when a young feather is fully grown it is sealed off from its blood supply and is essentially "*dead*". Scales are nothing like feathers.

Also, birds have special muscles and bones perfectly designed to help them fly. Some birds even have bones and muscles so perfectly made they are more agile and precise in flying than the most sophisticated modern fighter jet.

There are some scientists who imagine that all of the changes from dinosaurs to birds happened while the small dinosaurs were living in trees instead of on the ground. Either way, no one can explain how these animals could have survived with half-wings or partial feathers. No one can explain how they could have developed the special brain/nerve system that birds need in order to fly.

It must have taken many millions of years for these reptiles to have had enough offspring with longer and longer scales and higher and higher jumping abilities to become birds. If theropod dinosaurs gradually changed into birds over millions of years, there should be dozens of examples of in-between forms in the fossil record.

One of the most popular animals scientists have said is a transitional form between reptiles and birds is called *Archaeopteryx* (ark – ee – op – ter – ix). *Archaeopteryx* was a small bird about the size of modern day crow. It had small claws on its wings near the wrist joint, a long reptile-like tail and small teeth in its jaws. Modern birds do not have long reptile-like tails nor do they have teeth in their jaws. For these reasons, *Archaeopteryx* is thought by some Evolutionists, but certainly not all of them, to be a transitional form.

You can still see this animal in many public school textbooks. There are, of course, a great many reasons not to think that *Archaeopteryx* is an animal in-between reptiles and birds.

For example, the feathers of *Archaeopteryx* are 100% bird feathers. They are perfectly made for flying. Just like birds today, the quill of the feather fits toward the front edge of the feather. Under a microscope, *Archaeopteryx* feathers are built with tiny "*hooks and eyes*", just like modern feathers. The wings are just like a modern bird wing too. There is no hint of scales becoming feathers or evidence of the wing changing from the front arm of a theropod dinosaur.

The claws on *Archaeopteryx's* wings are nothing special either. There are three living species of bird today which have claws on their wings. These are the *hoatzin*, the *touraco* and the *ostrich*. Claws on the wings do not prove anything about dinosaurs becoming birds.

The teeth in the *Archaeopteryx* mouth are unserrated, small teeth. These teeth do not really prove much since there are other fossil birds that had teeth in their jaws. This fact only proves there are no birds with teeth in their jaws living today. We have to remember that hundreds of species or varieties of animal and plant life go extinct every year. We should expect to find all kinds of interesting animals that no longer live in our modern world. Of course teeth don't really tell us much anyway since there are reptiles living today that do not have teeth at all.

Besides, just because extinct birds had teeth does not prove anything. After all, there are animals living today that share features and they are not considered in-between forms. For example, the platypus is a mammal that lays eggs but no one thinks the platypus is either becoming a bird or changing from a bird into a mammal! Actually, platypus eggs are more similar to turtle eggs, which is even more bizarre.

Some Evolutionists think *Archaeopteryx* must have been an in-between form because it didn't have a breast bone. Flight muscles are attached to a bird's breast bone in living birds. But *Archaeopteryx* did have an extra-large wishbone (called the furcula), and only birds have a wishbone. In some cases, flight muscles may be attached to the wishbone instead of the breastbone.

Finally, some scientists have found undoubted bird fossils in the same or older-age rocks as *Archaeopteryx*. How could *Archaeopteryx* be a missing link between reptiles and birds if true birds lived right alongside it or were even older? The truth is, *Archaeopteryx* was a true bird[115] and none of the fossils scientist think might show dinosaurs changing into birds is convincing.

How to remember this:

Dinosaurs became birds? Don't be ab-su-r-d!

AB.-SU.-R.-D.

- **AB** = ARCHAEOPTERYX was a true BIRD.
- **SU** = SCALES and feathers are completely UNALIKE.
- **R** = ROCKS older than Archaeopteryx have bird fossils.
- **D** = DETAILS like claws on the wing are found in three living bird species today.

Question 15

Horses: What's with the chart in my textbook that shows the Evolution of the horse?

Many high school and college biology textbooks continue to show the fossil horse series as evidence of Evolution. The fossil horse series begins with a drawing of the leg bones of *Hyrocotherium* (Hi – row – co – theory – um) (or *Eohippus*), a three-toed animal. Included is a picture of how the animal may have appeared in life.

Next in the series is *Miohippus* (my – o – hip – us), which also had three toes but two of these toes are small. Then there is *Merychippus* (Mary – chip – us), an animal with one large main toe and two very small, apparently useless toes on the side of the main toe. Finally, the modern horse (*Equus*), with one toe, is shown as the result of this supposed long, slow change from three toes to one. Each of these animals is also shown growing in size and horse like appearance. The entire plot appears to be an open-and-shut case for the theory of Evolution – but there are some serious problems with this picture.

In the first place, the fossil bones shown in the pictures are for the rear feet only. No one says anything about the front feet of these animals. For example, the first animal in the series, *Hyrocotherium*, actually had four toes on its front feet and three on the back feet. *Hyrocotherium* was actually almost identical to the modern *hyrax* and is only included in the series because they needed to start somewhere. *Hyrocotherium* was chosen as the starting point but its habitat and skeleton show that it was not a horse-like creature at all. In fact, *Hyrocotherium* was more similar to a rhinoceros than a horse.[116]

There is no evidence of any gradual change from the four toes on the *front* feet of *Hyrocotherium* to the one toe on the *front* feet of *Equus*. Since only the rear feet are supposed to show some kind of smooth Evolutionary change, Evolutionists must conclude the front feet of these animals made a series of sudden jumps.

This group of animals also doesn't show any consistent change with the number of rib pairs they had either. *Hyrocotherium* had 18 rib pairs, *Miohippus* had 15, *Merychippus* added a few for 19 and the modern horse has 18. There is no clear Evolutionary series in rib pairs in this group.

Another problem with this series has to do with the fossil teeth of these animals. *Miohippus* had browsing teeth while *Merychippus* had high-crowned grazing teeth. There is no evidence of slow and gradual change in the teeth of these animals.

These animals were also specifically picked out of the fossil record, no matter where on the Earth they were found, to try and show this supposed change from three toes to one. Some of the fossils are found in North America, some in Europe, some in India and others in South America. Clearly, Evolutionists can't just pick and choose which fossils they want for their series without regard to where they came from. They must be able to show a steadily changing population in order to prove Evolution has occurred. If they can just pick and choose, they could line up some bones from China, a few from North Africa, a few from Antarctica and a few from New York City to prove there is a steady progression from butterflies to pelicans. Such a large sample from vastly different areas makes it appear very unlikely these animals were all part of the same changing population.

The most serious problem, however, is that the fossils of modern horses have been found from the same rock layer as *Hyrocotherium*. In fact, the majority of fossils from this supposed Evolutionary change came from the same fossil layers. This is simply not possible if the *Hyrocotherium* population was gradually changing into the modern horse. The original *Hyrocotherium* should have died out millions of years earlier as the modern horse emerged.

Dr. Duane Gish, a very well-known Creationist, pointed out that South American horses are found in reverse order. The one-toed variety (*Thoatherium*) is buried in deeper sediment layers than the three-toed *Macravchenia* which would make them older from an Evolutionist's point-of-view. The *Macravchenia* also lived at the same time as the South American one toed animal *Diadiaphorus* which also had two small side toes similar to *Merychippus*.[117]

Many Evolutionists have quietly agreed that the fossil horse tree is not real evidence of Evolution. Many Evolutionists today now believe these animals most likely existed in a large community of animals, much like animals on the African savanna today. They now consider horse Evolution "*jerky*" or "*patchy*" and reject the horse series as it was once so proudly shown. They recognize these animals are not a group of transitional, or in-between forms.[118]

How to remember this stuff:

In many a textbook you can find the HORSE chart, the thing to remember is it's FALSE from the START.

Horse / False / Start

- **H** = HORSE Evolution charts only show the back feet.
- **F** = FOSSILS of horses are found in the same layers as the first animal in the chart.
- **S** = SMOOTH changes are not found in rib pairs and teeth.

Question 16

Zoology: Did modern animals come from the survivors on Noah's ark?

There is an amazing ability in life-forms to produce a huge range of variety in their offspring. Variety is what allows life-forms to survive in different environments. No one has ever seen variety cause one kind of life-form to change into a completely different kind of life-form. Still, variation within a life-form kind does show us that the Creator is an amazing artist.

In humans, we can see many varieties of skin colors, height, face shapes and types of hair, among other things. There is a stunning ability for human beings to produce a huge number of varieties.[119] This is called "*variability*". In fact, the DNA codes from just one human male and one human female could produce 10^{2017} varieties. This is a 1 followed by 2,017 zeros. This number is larger than the number of stars in the sky.[120] This means there is more than enough variability in one human male and female to eventually produce every variation of human in the world today.

The same is true in animals. They have a range of variability that allows their kind to survive in a wide range of environments. God created the Earth's weather patterns to change from time-to-time which means that life-forms need to be able to adapt as the world changes around them. If this were not so, an animal population would either have to travel to a better area or risk dying out entirely.

Sometimes, a life-form group can become trapped in an area for a long time. In this case, the variability of the group is limited. This is called having a "*limited gene pool*". If the environment changes, one type of variety may end up standing out. In humans, wide land barriers and differences in cultures caused isolated groups where the DNA variability was very limited. Over time, traits such as height were able to stand out. For example, humans from a Pygmy tribe can be, on average, around four feet tall, whereas other tribes may average seven feet tall.

Skin color is an obvious variation in humans as is skull shape. But all of these varieties of human are still 100% human. All humans share the same capacity for intelligence and giftedness and are not more or less equal than any other variety. There is, in fact, only one race – the human race. All human varieties are able to mate with one another and all humans share a basic human nature too. We all create both the good and the bad: war but also art, slavery but also music, racism and science, lies and language.

"Animals today, from the ark made their way"

In animals, variability can also have important results that go far beyond skin color. For example, different varieties of dogs have different coats, heights, colors and habits. All dog varieties can mate with each other, but certain varieties of dog (*Canis lupus familiaris*) will have the ability to survive better in some places than in others. For example, the wolf (*Canis lupus*) tends to prefer colder climates. It has the body shape, coat thickness and social abilities to survive in the Arctic while its close cousin, the coyote (*Canis latrans*) traveled to warmer climates and led a more solitary lifestyle. This is not Evolution but variation within a life-form kind.

Some Evolutionists argue that creatures like the marsupials - such as kangaroos and koalas - are evidence for Evolution. After all, how could these creatures exist mostly in Australia unless they had evolved there? These animals are indeed very special but are they evidence of Evolution?

It is true that there are no natural populations of kangaroos in America or Europe, but this does not mean they must have evolved in Australia. There are, for example, marsupial opossums naturally occurring in North America. Of course, what makes marsupials different from other mammals is how they reproduce. Marsupials give birth to very premature offspring. The tiny offspring crawls into the mother marsupial's pouch and develops there. It is very, very unlikely that the way marsupials reproduce came about by accident in two different populations of mammals in two different places in the world. It is much more reasonable to think that kangaroos, koalas and other marsupials migrated to Australia, and were later cut off from the rest of Asia by rising ocean levels. Since they have been isolated there ever since, they became the main type of mammal in Australia. But how did they get there?

Biblical Creationists believe there was only one ice age, an event triggered as an "*after-effect*" of Noah's Flood. The ice age would have lowered sea levels all over the world, creating land bridges between Asia and Australia. Animal movements right after the Flood were probably very aggressive as some animals moved to escape the northern cold and others moved quickly to fill up places in new environments. Later, when the ice age ended, the ice melted, ocean levels rose and some animals became isolated. Years of mating in these smaller gene pools allowed certain strong traits to rise to the surface.

It is possible for every race of humans and every variety of animal-kind represented today to have descended from the original pair that went into Noah's ark.[121]

How to remember this:

Animals today, from the ARK made their way!

A.R.K.

- **A** = Animals have a great ability to have many varieties.
- **R** = Restricted movements after ocean levels rose isolated certain animals.
- **K** = Kinds of animals will never change into completely different kinds.

Question 17

The Missing Link: How do Evolutionists think people became people?

People are very interested in finding out how the human race became human in the first place. Learning how the laws of chemistry say life is not a chemical accident is not as personal as learning about where humans came from. This means that fossils of "*human ancestors*" are usually front-page news. People everywhere are both interested and sensitive about this issue. As a result, there have been quite a few twisted facts and even outright lies from the experts, right from the beginning of the "*Evolution Revolution*" that began in 1859.[122]

Evolutionists believe that humans (*Homo sapiens*) evolved from an ape-like ancestor that was also the ancestor for today's apes and monkeys. This ancestor began to evolve into both apes and humans sometime between 3 and 6 million years ago. The common ancestor for both apes and humans came from the prosimians, which were lemur-like animals. The prosimians supposedly evolved from small insect-eating mammals like the tree shrew which is a small mouse. This entire idea is total guess-work. Anthropologists are scientists who study humans and human ancestors and most of them do not agree on exactly how they think humans became human.

The most popular story for human evolution starts with an animal called *Australopithecus afarensis* (Awe - stral - o - pitch - icus Af - ar - in - sis). Many scientists think the *Australopithecines* evolved into various "*ape-man*" species including *Homo habilis, Homo erectus, Homo sapiens neanderthalensis* and finally, modern humans. This family tree is based on very little real evidence. In fact, most of the evidence is just fossil teeth because the rest of the body is usually not found.

Scientists now say that many animal fossils once thought to be human ancestors are not human ancestors after all. Animals like *Ramapithecus* (now called *Sivapithecus*), *Zinjanthropus boisei* (now called *Paranthropus boisei*), *Gigantopithecus blacki* and others, were once put into textbooks as facts about human Evolution. After a few decades, they became unpopular or were reclassified as something else. The picture from the fossils of how humans supposedly evolved is not clear to anyone.

Today, people often think that a savage ape-man ancestor for humans is an obvious fact. The truth is, fossil evidence for supposed human ancestors is regularly re-evaluated, changed, re-dated or thrown out altogether. This has been true of every supposed ancestor for man ever since Charles Darwin first published his book about Evolution in 1859. Today, Evolutionists are still arguing about the *Australopithicines* but they are the most popular *possible* human ancestor.

It is only fair to point out that new information can change what we think we know. This is what science is supposed to be all about. Scientists study, conduct experiments, argue and re-evaluate what they learn all the time. If scientists can change their minds when they learn something new, then they can avoid making mistakes. This is a good thing. The problem is that scientists can also be so committed to an idea or hypothesis, they can miss the bigger picture too. Ideas about human Evolution are a good example of this problem in thinking. Arguments scientists make about human ancestors never question the bigger picture of Evolution as a whole. Evolution itself is always assumed to be true, no matter what the evidence says, and it is only the *order* of which ape-like ancestor that came first that is being questioned. No one seems to ask the obvious question: if there is no real evidence of any in-between or transitional forms between any of the major life-form kinds, then Evolution as a whole is simply not true!

Many Evolutionists claim there are not enough fossils to really see how humans evolved. But this isn't really true either. Fossils that scientists think are human or ancestors of humans are called "*hominid*" fossils and there are actually more than 6,000 fossil individuals that scientists can study. This means the fossil record of hominids is actually very rich. However, most of these fossils are either too modern looking to help Evolutionists prove their idea or they don't fit very well into the Evolution model.[123]

Of course the number of fossil hominids Evolutionists think belong to our family tree is much smaller because they are very carefully selected. In fact, it would be easy to fit all the evidence that Evolutionists actually use to support human Evolution on top of your average dining room table.

Homo erectus

Since very few hominid fossils are thought to be part of our family tree, whenever a new one is announced, there is a lot of excitement and interest. A new "*missing link*" in our own history is major news that brings glory and fame to the lucky fossil hunter. Not a few scientists have made big claims for fame when the evidence is actually unclear, misleading or an outright lie.

A famous example is called Java Man, a skull and leg bone fossil that scientists now call *Homo erectus*. Eugene Dubois, a medical doctor from Holland, discovered these fossils in 1891 on the island of Java. He named it "*Pithecanthropus erectus*" or "*erect ape-man*". These two bones became the basis of the entire *Homo erectus* species.[124] Many fossil remains assigned to *Homo erectus* are only teeth, many are probably apes and others are clearly outright fraud. It was shown later that Java Man, for example, was not what he appeared to be and that Eugene Dubois was behind the deception.

Dubois failed to mention that two fully human skulls were found in the same area and at about the same depth as the Java Man fossils. This is a major piece of evidence that Dubois concealed for thirty years. He simply took evidence from a site that matched the idea he wanted to support, then ignored and suppressed the evidence of modern humans at the same site.[125] Humans obviously did not evolve from "*Java Man*", they existed within a similar time frame with "*Java Man*".

Unfortunately, scientists continue to include a number of very questionable fossils in the *Homo erectus* group. It's actually easy to see that most *Homo erectus* fossils are either true human fossils, a mixture of several animal remains or types of extinct ape.

It's actually very difficult to say exactly what *Homo erectus* really is in the first place. Scientists don't agree. For the most part, *Homo erectus* is supposed to have had a smaller average brain capacity than modern humans and some skeletal features in common with Neanderthal man. The *Homo erectus* skull is usually long with low brow ridges and little or no chin. From the neck down, undoubted *Homo erectus* fossils are identical to modern humans. For example, KNM-WT 15000, a fossil *Homo erectus* skeleton of a 12 - 13 year-old male, was 5 feet 6 inches tall and would have stood around 6 feet when full grown.

Peking Man

Another highly suspicious *Homo erectus* group of fossils was discovered near Peking, China in the 1920s. The fossils were originally called "*Sinanthropus pekinensis*" but today modern scientists consider the "*Peking Man*" remains to be *Homo erectus*. About 30 skulls, all very broken, various fragments of animal bones and teeth were found at the site.

All of the Peking Man evidence, except two teeth, was lost during World War II. The original 30 skulls were supposed to be very ape-like and the find seemed like an important boost for the theory that man's ancestors were ape-like from the neck up. After all, the Peking Man site had at least 30 skulls to examine.

All that remains today are plaster casts of the skull fragments, but some important evidence at the site was overlooked. For example, six skeletons of modern humans were also discovered at the site. There was also evidence of limestone quarry work, fire pits and shelter foundations. Some scientists wonder if the original fossils were really *Homo erectus* after all.

All of the original 30 skulls show evidence of being smashed by some type of club. No bones below the neck were found except, of course, the six skeletons of modern humans. Many of the skull pieces were also burned. It is possible that these skulls are not human ancestors at all. Perhaps ancient quarry-workers killed large monkeys for food and smashed open the skulls to eat the brains. The smashed skulls are all that is left of this ancient meal.[126]

The majority of fossils called *Homo erectus* have an average brain capacity of 700 - 1250 c.c. This is within the range of modern humans who have a brain capacity from about 700 - 2200 c.c. Most of the dates assigned to *Homo erectus* bones, on an Evolutionary time scale, are within the time frame of modern humans. If scientists take out suspicious material like Peking Man and Java Man, then *Homo erectus* and Neanderthal Man are virtually the same. Only a belief in Evolution causes scientists to call these people a name other than "*human*" even though the only real difference between modern humans and *Homo erectus* is skull shape.[127]

Piltdown man

Piltdown Man was a fossil skull and jawbone discovered around 1912 near Piltdown, Sussex, England. The skull was very man-like. It seemed to have a modern brain size but the jaw was very ape-like. Some of the world's greatest anatomy experts said that Piltdown Man was "*the missing link*" between apes and humans. They called it "*Eoanthropus dawsoni*", and remained convinced it was real for 40 years. In 1953, however, Piltdown Man was discovered to be a fake. Someone had taken a modern ape jaw and human skull, painted them to look old and filed down the teeth to make the pieces fit. This fake fooled the world's experts for 40 years.[128]

Nebraska Man

Nebraska Man was built around a single tooth discovered in Nebraska in 1922. It was given the scientific name "*Hesperopithecus harolodcookii*". An artist's reconstruction of Nebraska Man was published in the *Illustrated London News.* The famous and respected paleontologist, Henry Osborn, supported this find, but in 1927 - 1928, everyone realized a big mistake had been made. It turned out that the tooth had actually come from a small pig.[129]

"If we exclude the possibility of Creation, then obviously man must have evolved from an ape-like creature, but if he did, there is no evidence for it in the fossil record."[130]

How to remember this:

Human fossils are very odd, many fakes and many frauds, and fossil HELP is only found, when wishful thinking jumps around.

H.E.L.P.

- **H** = **H**OMO ERECTUS does not affect us, he had brains for all to see.
- **E** = **E**UGENE'S Java man was found, with human fossils all around.
- **L** = **L**ET down for Nebraska man, he turned out to be a pig.
- **P** = **P**ILTDOWN man was just a fake, a skull, a jaw and dye to make.

Question 18

Australopithecus: Who was Lucy?

Raymond Dart discovered a group of ape-like animals in South Africa in the 1920s. They were called "*Australopithecus*" (Awe - stral - o - pith - icus). The teeth in these animals are similar to human teeth, so some scientists began to think they were in the human family tree. Since that time, more fossils of these animals have been found and studied.

The most popular idea among Evolutionists today is that *Australopithecus* is a direct human ancestor, but not all scientists agree. Bones that scientists think may be human ancestors are called "*hominid fossils*". Many hominid fossils that scientists think are *Australopithecus* may not be at all. Some fossils are only called *Australopithecus* because of *where* the bones were found, not because the bones really tell us anything.

There was a larger type of this animal called A. *robustus* (also called *Paranthropus robustus*). It was about the size and weight of the modern gorilla. It had massive jaws, teeth and a bony ridge on its skull called a sagittal crest. The smaller variety is called A. *afarensis*. It was about the size and weight of a modern chimpanzee. Both of these animals had small brains about the same size as modern apes. In fact, the only truly different thing about *Australopithecus* was a few minor features of their teeth. The truth is *Australopithecus* was probably just an ape.

A. *afarensis* had front teeth that were fairly small compared to its cheek teeth. The cheek teeth were about as big and thick as the cheek teeth of a 400- pound gorilla but A. *afarensis* only weighed about 60 pounds. Evolutionists think this is evidence of an ape-like creature changing slowly into a human because humans have much smaller teeth than apes.[131]

Some scientists think A. *afarensis* may have walked upright like humans, at least some of the time. Modern chimpanzees will also walk on two legs from time-to-time but no one thinks this is evidence they are becoming human today. The evidence that these animals walked like humans today is very weak. The idea that A. *afarensis* walked like humans do came mostly from a small female A. *afarensis* found in 1974 in Africa that scientists named "*Lucy*".

In 1974, Donald Johanson and his team discovered Lucy near Hadar, Ethiopia. Lucy's skeleton was about 40% complete. This was the most complete skeleton of an A. *afarensis* ever found. Johansen thought Lucy was about 3 million years old. Lucy was much older than expected for a human ancestor. Dr. Johanson said that Lucy was the oldest known direct ancestor of humans. He also shocked everyone when he said that Lucy once walked just like humans but was basically an ape from the neck up. In other words, Lucy was an ape-woman.[132]

This was big news. Many museums and science magazines supported the idea that Lucy was our direct ancestor. Many reconstructions of the fossils and artists' drawings showed Lucy walking like a human. She was shown with an intelligent face but having about half as much hair as a chimpanzee.

But not every scientist agreed with Johanson and his supporters. Lord Zuckerman, a famous British anatomy expert, and Dr. Oxnard, a professor at USC, each spoke out against the idea that Lucy walked upright like a human. They compared Lucy's bones with other animals and humans. They performed detailed computer work to study this idea. After over 15 years of in-depth study, they each decided not to support Dr. Johansen's ideas.[133] They decided A. *afarensis* was not our ancestor and did not walk upright as humans do today.

Also, the hand fossils of Lucy are curved and were heavily muscled, much like an orangutan's are today. The pelvis was balanced more like a chimpanzee's, too. Finally, studies on the knee joints of other *Australopithicines* showed they are within the range of spider monkey knee joints, not human knee joints. It seems that Lucy and others like her were very similar to modern chimpanzees and gorillas.[134]

Most importantly, there is strong evidence that modern humans existed at the same time as Lucy. Fossil bones such as KNM-ER 1470, KNM-ER 1481 and KNM-ER 1590 are identical to modern human bones. In fact, these fossils were described as completely human at first. Later, Evolutionists realized that the rock layers in which these human bones were found were dated the same as the rock layers where *Australopithecus* fossils were found. This caused a problem until scientists just decided to say these bones were not really human, just "*unusual*".[135] KNM-ER 1470, 1481 and 1590 are clearly human but they don't fit into the Evolution story.

There is more evidence that modern humans lived at the same time as Lucy. A trail of footprints is preserved in hardened volcanic ash in Laetoli, Africa. Evolutionists think this trail of footprints is 3.6 - 3.8 million years old. This is about the same age or older than the Lucy fossils (on an Evolutionist's time scale). Evolutionists think this shows *Australopithecus* animals walking like humans - but these prints were clearly made by humans. Actual scientific studies of the footprints of humans who never wear shoes, have shown that the Laetoli prints are exactly the same as human prints.[136]

The truth is, Lucy and others like her were very similar to modern chimpanzees. Since we find modern human bones from the same time period and human footprint trails, there is no reason to think Lucy was our ancestor.[137]

How to remember this stuff:

Lucy's a fossil we really should SKIP. Her feet were just hands that gave branches a grip.

S.K.I.P.

- **S** = SKELETON: Lucy's bones show she did not walk upright in human manner.
- **K**= KNM-ER 1470, 1481 & 1590: all fully human fossils found in the same rock strata as Australopithecus.
- **I** = IMPRESSIONS of fully human footprints have been found that are supposedly as old as Lucy.
- **P** = **the** PAWS (hands) of Lucy were curved and heavily muscled, designed for gripping branches as are the Orangutan's today.

Question 19

Cave Men: Who were the Neanderthal people?

Since scientists find stone tools and evidence that "*early*" man lived in caves, there is a common belief today that the "*first*" humans were savage cavemen. For many years, people thought that cavemen were primitive and could barely speak. Others think that cavemen were not that far from apes.

The most famous cavemen are the Neanderthal people. For more than a hundred years, scientists thought the Neanderthal people were an in-between form part-way between ape-like creatures and true humans. Neanderthals had heavy eyebrow ridges and low, sloping foreheads. They also had a thick jawbone with no real chin. Most of them were short in height and had bowed legs. Many scientists believed they walked in a shuffling, ape-like manner. Scientists described them as moving around with heads jutting forward, knees slightly bent and arms dangling at the side. Neanderthal Man was the perfect picture of the savage ape-man. Over the years, however, evidence has shown that the Neanderthal people walked just like we do today. They also had a spoken language and a very strong culture.

The first Neanderthal was found in 1857 at a limestone quarry in the Neander Valley, near Dusseldorf, Germany. Rudolf Virchow, a medical doctor on the scene, thought the skeleton belonged to a very old man with arthritis and rickets. Over time, so many skeletons were found with the same traits that Evolutionists decided the Neanderthals were a missing link between ape-like animals and modern humans.

Over 300 Neanderthal sites have been studied since 1857. We now know that Neanderthal was not an in-between form linking apes with humans. Instead, many Evolutionists think they were an offshoot of human evolution that came to a dead end. Creationists believe Neanderthal was a true human being and there is strong evidence to support this position.

On an Evolutionist's time scale, the first Neanderthals suddenly appeared out of nowhere about 200,000 years ago and then disappeared mysteriously about 35,000 years ago. Neanderthals also had big brains. In fact, the average brain size of a Neanderthal was larger than that of modern humans. This was a problem at first because Evolutionists thought brain size showed intelligence. After all, Neanderthals were supposed to be stupid and violent cavemen, so how could they have such big brains? But over the years, brain research has shown that brain size does not always tell how smart you are.

Neanderthal man was physically powerful, much stronger than modern humans. They were very religious too and buried their dead with flowers as we do to this day. Their culture was very complex. They were fully capable of language and cared for the old and crippled until they reached a natural end. If you gave a Neanderthal man a shave and dressed him in a suit you wouldn't notice any difference.[138]

Most scientists now think Neanderthals were humans but still think they were fairly primitive. They probably think this because many Neanderthals lived in caves and used stone tools. They were also very rugged-looking and had strangely shaped skulls but the Neanderthals obviously had brains.

The truth is, looks aren't everything and stone tools do not mean people are not smart but it's probably true that Neanderthals wouldn't win any beauty contests today. Still, Neanderthal looks can be explained and how they looked does not mean they were in-between apes and humans.

In the first place, many Neanderthals had problems with a disease called rickets which comes from a lack of vitamin D. Rickets softens bones causing them to deform.[139] There is also evidence showing that the environment during Neanderthal's lifetime was harsh. It is very possible they were isolated from other humans also. This would have caused certain types of traits to show up more often because of a smaller gene pool. These factors can explain why these people had different skeleton features than modern humans, so many scientists today think Neanderthals were fully human.[140]

It could be that people became cave-dwellers because they didn't have much choice. Right after Noah's Flood, the world fell into an ice age. In places like Egypt, where the climate was warmer, amazing cultures were able to come together. Other tribes of people were forced by war and fear into tough places closer to the great ice sheets in the north. People had to seek shelter as best they could and come up with different kinds of hunting techniques for survival. Lack of sunlight, poor diet, a smaller gene pool and an overall lack of vitamin D probably caused Neanderthals to look as they did. After these conditions changed, Neanderthal tribes quickly lost these skeleton traits. This idea can explain why they seem to suddenly show up and then mysteriously disappear. What we do know for sure is that Neanderthals were not "*ape-men*".[141]

How to remember this stuff

Neanderthal Man was a true HUMAN being. Studies have shown this; now scientists are agreeing.

H.U.M.A.N.

Neanderthal man was fully HUMAN.

- **H** = HEALTH was impaired by a lack of vitamin D which can cause bones to soften.
- **U** = USE of stone tools and caves were not a sign of lack of intelligence.
- **M** = MINIMAL gene pool allowed some physical features to stand out.
- **A** = AVERAGE brain size was larger than modern man.
- **N** = NOAH'S flood left a struggle to survive using whatever they could.

Question 20

The Beginning: What is the law of cause and effect?

The Law of Cause and Effect says that there cannot be an effect without a greater cause to produce it. In other words, nothing just happens unless something else causes it to happen. Mountains erode over time because of rain and wind and snow. The checking account balance is overdrawn because the money was already spent. Animals die because of old age and disease and the tides flow in and out because of the moon's gravitational pull. The list is endless because every effect has a force of some kind behind it. The universe itself is also based on the Law of Cause and Effect.

Knowing this leads us to just two choices about the universe. The first choice is to believe the universe has always existed. In other words, it had no beginning and will have no end. The second choice is to believe the universe had a beginning, a "*first cause*", that made all of the other things happen in a long, long chain of cause and effect.[142]

The first choice is difficult to believe. If the universe had no beginning and will have no end, then it should not be winding down and wearing out all around us. The Second Law of Thermodynamics helps explain why and how the universe is unwinding. The law describes energy like the ripples in the surface of a lake caused by a pebble thrown into the water. As the ripples spread out from the original source, they lose energy until they gradually fade away altogether. In the same way, the universe is rapidly wearing out. It has less usable energy every day. If the universe is eternal, it shouldn't be wearing out.

The second idea is that there must be some kind of super-powerful source of energy and information that started everything. This "*first cause*" must be bigger than time and space and matter and energy and everything else in the universe. The "*first cause*" must be able to create energy out of nothing. This idea makes the most sense because after studying the universe for many years, it is clear that nature is not eternal. Natural things must have had a beginning.[143]

The Bible teaches us that God created everything out of nothing, just by His own power. This means that God is the "*first cause*" of the universe. Of course, everyone knows that when an artist makes a painting, he puts a little of his own personality into what he creates. So, in the same way, we should be able to see something of what God is like from what He has made.

The universe is very, very complex. It is also very orderly and vast beyond our imaginations. It is incredibly beautiful and there are more stars in the night sky than grains of sand on all the world's beaches. There is love and joy in this Creation but anger and sadness too. There are cold winds and warm places where children learn and play and grow. There are silly things and serious things and things that make you cry when you're really actually very happy. It is a marvelous place full of wonder and mystery.

The designer of all these things must be greater than all of it combined. He must be far more complex, more majestic and more beautiful than all the stars in the sky. He must laugh louder, love better and feel sadness and anger more deeply than all the world. He must be the master of all the winds and the waves and the fires in the stars and He must know how all of it works.

It is easy to see that many religions do not describe a god who can make a universe like ours. Many religions are really just taking natural forces like the wind or fire and calling these forces some kind of god. Some religions create gods out of the torn up body parts of other gods. Others teach that the world has always existed or that man is god in some form or another. But none of these ideas describes a "*first cause*" who can create energy and order out of nothing. These religions do not match with what we actually see in the universe.

The true "*first cause*" of all things must be a single, unified being. We know this because wherever we look in the universe, there is only one type of engineering. There is only one type of complexity and there is one set of natural laws that makes everything work. Dr. Morris said this is why we call it the "*universe*" instead of a "*multiverse*".

The true "*first cause*" must also be limitless because space is limitless. He must also be eternal because He existed before time began. The "*first cause*" cannot be a human-like god either. After all, humans can't even keep this tiny planet in order. The "*first cause*" must be greater than human nature in order to govern the universe. He must also be patient enough to allow humans to have freedom - no matter how much evil we bring upon ourselves - instead of forcing us to serve Him.[144]

The Bible alone describes God this way. There is more evidence to support what the Bible teaches than any other book ever written in any language. The evidence of history, literature and science all tell us the Bible is worth our trust. The Bible says that God is eternal and that we cannot measure Him. It tells us He has all power so that He can create things out of nothing. It tells us He is everywhere, unseen but knowing everything. The Bible says God is NOT part of the Creation but instead, everything in the universe stays together by His command. It says God is alive and has deep feelings. It tells us that God created everything, including the first two human beings. This means He created you and me too.

"For since the creation of the world God's invisible qualities - his eternal power and divine nature - have been clearly seen, being understood from what has been made, so that men are without excuse." (Romans 1:20, NLT)

How to remember this stuff

A CAUSE is a trigger and an EFFECT will result, so God must be bigger than the stuff in the vault.

Question 21

The Bible: Why can I trust the Bible?

The laws of science and the evidence we study tells us that God really did create the universe. He also created the laws of nature that determine how things work and stay together. Of course, no one knows how the universe is put together better than the One who built it in the first place. Most people are willing to admit there must be a God, but why should people think the Bible is anything more than any other religious book? Shouldn't it be enough just to give in and admit there must be a God? After all, maybe all the world's religions are not wrong but just shades of the truth!

The problem with this idea is that the universe simply does not work this way. There are natural laws such as the Law of Gravity that are true whether we want them to be true or not. In the same way, not all religions can be true at the same time. C.S. Lewis explains this idea like this:

"*As in arithmetic - there is only one right answer to a sum (addition problem), and all other answers are wrong: but some of the wrong answers are much nearer being right than others.*"[145]

Many religions and religious books do have similar ideas. But as C.S. Lewis said, this is because some are closer to the right answer than others. But if you are trying to figure out a math problem, the only way to see if the answer is correct is to make sure we are doing each step in the problem the right way. In other words, we have to add and subtract using the rules for addition and subtraction or we are bound to get the wrong answer. We may get an answer that is close to being right, but close is still...wrong.

In the same way, it is important to see how the Bible fits with the nature of Creation. This confirms for us that we can trust what the Bible says about the Creator. In other words, if the Bible really fits in with what we know about God from the Creation, this would be like checking our math to see if we have the right answer or not. For example, the DNA code in living things provides planned, purposeful information. This means that living things each have a *purpose* - there is a *plan* for their structure and a *place* for them to live. It makes sense, then, to think that the Creator of DNA would also provide planned and purposeful information to explain the Creation to His creatures. The Bible fits with this idea because the Bible says, "*All Scripture is inspired by God and is useful to teach us what is true and to make us realize what is wrong in our lives. It straightens us out and teaches us to do what is right. 17 It is God's way of preparing us in every way, fully equipped for every good thing God wants us to do.*" (2 Timothy 3:16 – 17, NLT).[146]

The Bible fits the nature of Creation in more ways than any other book ever written in any language. For example, in the Creation we see something called "*genetic integrity*". This means that one life-form kind can have some variety but will always remain the same basic kind. In the same way, the Bible has an internal integrity. It talks about some of the most controversial ideas in all human history – yet the Bible has a single, unified message from beginning to end. There is a great deal of variety in its pages but the message does not change. There is history, poetry, song lyrics, sermons and prophecy but the message does not contradict itself anywhere. This is a reflection of the same sort of thing we see in genetic integrity.

The message of the Bible has also been helpful during every time period in history. It was not just useful for ancient people but remains a life-changing power in the 21st century. The message of the Bible has transformed lives for the better in every people group, in every environment on Earth. In the same way, people are able to use the natural cycles of rain, tides, plant and animal growth to help them survive everywhere and at every time in history. If a religious book is only useful for one group of people, in one culture and in one time period, it would not fit with the nature of created things, which are useful everywhere and for everyone.

The Bible teaches that God created the universe out of nothing. It does not tell clever stories about gods having fights that created the oceans. It does not teach incredible things like saying that the Earth sits on the back of a giant turtle swimming through outer space as some religions do. The Bible explains how the world was made in a way that fits with the facts of chemistry, physics, the fossil record and every other science. In fact, as we have seen in this book, the facts of real science support the Bible's story of Creation far better than any other scientific or religious idea.

In nature, tiny mistakes in the DNA code between one generation and the next are usually deadly. These mistakes, called "*mutations*", are weeded out and do not change the basic life-form kind into a different kind. If the mutations are too severe, they are weeded out very quickly. There is also variety within a kind but this variety only makes the kind more interesting and beautiful and never changes the basic kind. In the same way, the Bible has been translated into more languages than any other book in history. There are many versions of the Bible too but the message does not change. Over the centuries, mistakes in copying the Bible have been found and weeded out – just as we see in DNA. Severe mutations are caught and weeded out even faster.

The Bible stands alone. There is no other book, religious or not, that fits in with the nature of the Creation like the Bible. It is a book written over a 1,500-year time span, in 40 generations, by more than 40 authors from every walk of life. It was written on three continents and in three languages: Hebrew, Aramaic and

Greek. But even written in such different places and languages, it has a powerful and single message. It teaches the answers to the most common questions people have about life but it never makes a mistake.

The Bible has also been read by more people and published in more languages than any other book ever written in any language. It is the best-selling book in world history. More people have tried to destroy the Bible than any other book but it has survived every attack. This fits in with the nature of Creation too because we see many environmental disasters but life always survives.

In nature, we can check the fossil record to see what life-forms looked like in the past. The evidence shows us that the major life-form kinds have a great deal of variety but have never changed from one kind into another. In the same way, we can check ancient copies of the Bible to see if it has stayed the same over time too. What we find is that there are more ancient copies to look at than there are ancient copies of any other book ever written. No other book can be checked as closely as the Bible.

For example, there are more than 25,000 ancient copies of the words of the New Testament alone. The New Testament is the second half of the Bible that was written just after Jesus lived on the Earth about 2,000 years ago. These copies show us that the Bible we have today is very, very accurate. We also have many outside witnesses to what the Bible says as well. These outside witnesses are writers of other books from that time period that show the Bible was telling what really happened. There are more of these outside witnesses for the Bible than there are outside witnesses for any other book ever written, anywhere, at any time.[147]

Even more incredible, the Bible tells the future and has never been shown wrong. Experts can prove these predictions were written hundreds of years before the events happened and they have come true 100% of the time. No other book or person can say this.[148]

The Bible is absolutely unique. This book, more than any other, fits the nature of Creation. We know that when an artist makes a painting, he puts some of his own personality into his work. In the same way, the God who created the heavens and the Earth put a little of His own personality into His work. This is why the Bible fits with the nature of the Creation because the Bible was also created by God.

How to remember this:

The b-i-b-l-e, that's the BooK For Me.

B.ooK. F.or M.e

- **B. K.** = BEFORE the events it was written and then it came true, KEY truths about God in its words we review.
- **F. M.** = FITTING in with all science and history throughout, the MANUSCRIPT copies we can always check out.

Question 22

Jesus Christ: What is a Christian?

"So God created people in his own image; God patterned them after himself; male and female he created them." (Genesis 1:27, NLT)

Just like a mirror reflects the image of a person looking into it, human beings are designed to reflect God's image. We should be able to look at a person and say, "*That is what God is like*". This means that everything we do or say or think should look just like the things that God says or does or thinks.

But of course, we don't look like this today. Whenever we do things that do not reflect the image of God, we betray what we were made to be. All of us have done things God would not do. Everybody says things God would not say.[149] The Bible calls this "*sin*". Too many people think that sin is only big evil deeds like stealing or murder. But the truth is, sin equals anything, even small things, which twist or distort the image of God in people.

No one has to learn how to sin either. No one had to teach us how to lie when we were kids. In fact, most parents have to work hard to teach kids how to tell the truth. No one has to teach us how to be self-centered and this is why we have to *learn* how to share our toys. Sin is something we have from birth and it all started with the first humans that God made – Adam and Eve. God made them to reflect His image, but they decided to do their own thing and go their own way.[150]

Some people like to blame God for all of their problems. They say God can't really exist, because there are so many bad things in the world. These people do not realize that the evil we see in the world happens because people do not act as God acts. People who do not reflect God as they should cause the evil we see in the world today. This is not God's fault; it is *our* fault.

God allows people to act this way for a reason - because He loves people and He wants us to love Him back. But love is a choice. God did not create us like robots. He built us to make choices so that we could love God in a real way. God hates the evil things we do to each other but in order to give people a chance to choose to love God, He had to allow us to decide not to love Him too.

The Bible tells us that our sins separate us from God. This is because anything that does not reflect God's image is pushed away from Him because of His purity and perfection. It is something like light and darkness. If it is dark outside your room and you open the door, the darkness does not bleed into the room. Instead, the light pushes the darkness away. In the same way, our sins are like darkness and God's presence is like light - we simply cannot be in the same room together.

This means that sin separates us from God. The Bible calls this "*death*" because God is the creator of life.[151] If we are separated from the Maker of Life, we are really dead - even if our heart is still beating. In the end, our bodies will die but our soul will continue to exist. If we are separated from God, this existence will be in the darkness, away from God. The Bible calls this "*the second death*".[152]

Another way to think of it is to see sin like spiritual AIDS. If you had AIDS, no matter how many vitamins you might take, it won't cure the problem. You can work out at the gym every day and be as muscle-bound and beautiful as a model, but it still won't cure AIDS. It will destroy you in the end. Sin is the same way - no matter how many "*good deeds*" you might do, no matter how hard you try, the sin disease is incurable and deadly.[153]

Without God's help, humans are lost and separated from God because of sin. The good news is that God loved people so much, He made a way for them to be saved from this separation from God. The Bible says that God Himself became a human being in the person of Jesus of Nazareth. Jesus had a human mother but no human father. This meant that the sin disease was not passed down to Him. He was 100% human and 100% God at the same time. Here's the exciting part: sin causes death but Jesus had never sinned. When Jesus died on the cross, He literally caused death to work backwards. Jesus rose from the dead! Jesus paid the price for sin, once and for all.[154]

The Bible teaches us that anyone who is willing to turn around from a sin-directed life and start a Jesus-directed life can be saved from sin. Jesus is willing to cover anyone with His own blood because His blood can wash away any sin. We can't do it ourselves, no matter how hard we try, but He can.[155]

In order to understand how this works, imagine you had committed a terrible murder. You are found guilty by the court and sentenced to death. There is no way out of this and no appeal. You have a serious problem!

What if the judge, who is completely innocent of the crime, took off his robe, came down from the bench, and said to you, "*I love you so much I will let the jailers execute me instead of you. I'll die in your place. But you have to accept my sacrifice or you will still be guilty and have to face the death penalty*". Since the penalty for the crime is death, if you accept his offer, then you would be legally free to go. The judge himself has paid the penalty for your crime.[156]

God has done just this for you and me. He paid the price we could never pay for sin through the death and resurrection of Jesus. If you accept this from your heart, then God will say that your sins have been paid for.[157]

Christians are people who decide to accept the judge's offer to die in their place. It is not difficult to do this – but you have to be real about your decision. You must first admit to God that you are a sinner and you need Jesus to pay for your sins. When you do this, you are doing two things: 1) realizing God has the right to say what is right and wrong, and (2) admitting you need to live His way.[158]

Next, you need to talk to God in prayer and ask Him to forgive you for your sins. Use the same kind of faith it takes to believe in Creation. You have to decide to believe, and you have to decide to put God in charge of your life.

That's it! You don't have to shave your head, wear a blue robe, eat only grass seeds on Saturdays, and dance on a street corner to impress God. If you believe and pray, God will forgive your sins and take charge of your life.

Christians will continue to struggle with sin. Being a Christian does not magically make you perfect on the spot. It does mean owning up to your actions that do not reflect God's image in prayer and continuing to try to live a life that does reflect God's image.[159] Things can be just as difficult for a Christian as for anyone else. But if you put all your trust and hope in Jesus, you can know for certain that you will be with God forever at the end of this life.

You can begin this new life following Jesus by praying a simple prayer. Remember, God will decide whether or not you are sincere. A foolish person will pray to God for forgiveness but then make no effort to follow God's way of life. God will not listen to that kind of prayer. You must mean what you pray. You can pray something like this:

"Father God, I pray to You in Jesus' name. I admit that I am a sinner. I have said and done things that do not reflect Your image and I am guilty. I have now come to understand that You made everything and you have the

right to make the rules in life. I believe that You became a human being in the person of Jesus. I believe Jesus died on the cross to pay the price for my sins and I believe Jesus rose again from the dead. Based on Jesus dying for my sins, I ask You to forgive me for my sins. I decide right now to turn around from a self-directed life and I make You the Master of my life. Amen."[160]

A person who accepts Jesus as the Captain of their life will begin to live life differently. It's a little like joining the U.S. Military: in the military, you have to turn away from a civilian life, then swear to follow the officer's orders and then go into training to live out the military lifestyle of service. In the same way, Christians are people who turn away from living just any way they want (repentance), then swear to follow Jesus as their Captain (pray and accept Him as master of all life choices) and then begin to train in living the Christian life (learning from other Christians and from the Bible – this is what church should be all about).[161]

People will know whether you are really a Christian or not by the things you do. God will know whether you were really sincere when you prayed to Him too. You can't fake it with God. He expects you to put all your trust in what Jesus did for you and you can't do this unless you begin a lifestyle of turning away from sin. The good news is that Jesus will carry you through to the end.

How to remember this:

Romans 10:13 (KJV) For "*whoever calls on the name of the Lord (Jesus C.H.R.I.S.T.) shall be saved*".

C.H.R.I.S.T.

- **C** = CREATION: We are created in the image of God but by our own choices we betray what we were made to be.
- **H** = HUMAN FORM: God became a human being in the person of Jesus Christ, 100% God and 100% man.
- **R** = RESTITUTION: Jesus paid the penalty for our sins by dying on the cross.
- **I** = IMMORTAL: Jesus rose from the dead on the third day. Anyone who turns from their own way and asks Jesus to forgive their sins, because of His resurrection, will be made right with God.
- **S** = SCRIPTURE: The Bible alone is God's teaching to us.
- **T** = TRIAL: Jesus is coming again and will judge each person according to what they have done. Those who put their trust in Christ will be forgiven and remain with God forever. Those who have not will be found guilty and remain separated from God in outer darkness forever.

Question 23

Now What? How can I use this information, especially in school?

We need to remember that we can deliver the right message in the wrong way. Even if the message you want to deliver is absolutely the truth, if you deliver it with a bazooka, there is going to be some damage. This is why Jesus taught us to treat others the way we want to be treated. This is the called "*the Golden Rule*" and we need to live by it even if other people don't. Just because many Evolutionists make fun of Creationists does not mean we should act the same way.

We need to keep in mind that scientists and other people who believe in Evolution are not stupid. Some of the most important scientific discoveries have been made by people who do not believe in God. There are many reasons people have for not believing in the Bible. Of course, if they really look at the facts of science carefully, they will see Evolution cannot be what really happened in the past.

Many Evolutionists are very sincere and dedicated professionals. They simply do not know the truth because they have been deceived. It is true that we should never accept Evolution as the truth and we should also be ready and willing to stand up for what the Bible teaches. On the other hand, we should not be rude or mean-spirited either.

If you find a chance to talk with someone about what you believe, make every effort to be respectful and kind. This is what Jesus taught us to do. The best way to talk about why you think Evolution is not true, is to first remember what James taught us:

"*My dear brothers and sisters, be quick to listen, slow to speak, and slow to get angry.*" (James 1:19, NLT)

An excellent way to be "*quick to listen, slow to speak, and slow to get angry*" is to allow the other person to speak first. Ask honest questions to see why the person believes Evolution is true. After a few minutes, ask if you can share why you think Creation is true. Most people want to be seen as tolerant. If they begin to argue, you can say, "*I listened to you without arguing or putting you down so I could understand you. Wouldn't it be fair for you to do the same for me?*" In most cases, people will take the time to listen to what you have to say if you respect them enough to listen first. In this way, you can "*be quick to listen*" and let the facts speak for themselves.

It is also important to remember that winning an argument is not how people come to believe in Jesus. In fact, Jesus said, "*...people can't come to me unless the Father who sent me draws them to me.*" (John 6:44, NLT) It is God who changes a person's heart. When we talk about Evolution, we are breaking down a wall that can keep people from listening to God drawing them. In the end, the best evidence in the world can still be ignored. It is not your job to convince anyone but it is your job to "*worship Christ as Lord of your life. And if you are asked about your Christian hope, always be ready to explain it. 16 But you must do this in a gentle and respectful way.*" (1 Peter 3:15 – 16b, NLT)

If you are a student in a public school, being able to answer for your belief in the Bible "*with gentleness and respect*" is especially important. You need to remember that all of the subjects taught in the school are part of the curriculum plan. This curriculum is put together by teachers, the school board, the state and the government and there certainly has been a great deal of argument about what should be in the science curriculum. For some years now, Evolution has been part of the curriculum but Creation has not. This is because many people see Creation as a religious opinion.

This doesn't seem fair, but it really could be an advantage. Think about it like this: Imagine you are going into battle. You and your army are armed with bows and arrows and swords. Your opponent, on the other hand, has an army armed with bombs. Who do you think is going to win? What can you do to even the odds?

Obviously, you have to learn all you can about your opponent's weapons. In fact, if you can find out about the problems in their weapons, you might be able to disarm them before they blow you up!

Or imagine it this way: If you have to take a huge exam about Greek myths, you would need to study and learn all you can about Greek myths. If you don't learn about Greek myths, you won't be able to answer questions about them. You will fail the test.

In the same way, the facts of science show that Evolution is a myth. Still, we may be required to know the story and answer questions about it in school. But just understanding a subject does not mean you have to believe it is true. Besides, you can't expect people to listen to you with respect if you won't take the time to understand their point-of-view too.

One of the most powerful bombs against believing in God is the myth of Evolution. Believing this myth can blow up a person's belief in God. It causes doubt and confusion. The best way to disarm it before it blows up and causes this doubt and confusion is to understand how it is put together. When you see that the facts of science do not support Evolution, it doesn't blow up your belief in God. For this reason, do your best on tests about Evolution but learn all you can about why it is only a myth.

But don't forget the Golden Rule! If you argue with teachers in science class instead of respectfully learning what they are teaching, you will only make people mad. If, on the other hand, you are the best student in the class, then a teacher may have enough respect for you to let you explain why you think Evolution is not true. If you are not disruptive in class and you know the story of Evolution better than everyone else, people in class will ask you questions. Remember, the Bible teaches us to love people - even our enemies.

"How NOT to deliver the Bible's message to your friends"

Use a little strategy when you ask questions about Evolution. Ask questions that make others look at Evolution very closely. Don't ask questions that are designed to embarrass people or make them look stupid. Instead, when you are taught that humans came from an ape-like animal millions of years ago, learn everything your teacher has to say first. When the teacher is satisfied that you know what the textbook says, then go ahead and ask questions that cause people to really look at the story. You could say, "*I've read that Lord Zuckerman studied this Lucy fossil for over 20 years and decided she didn't walk on two feet like people. Why then should we think this is our ancestor?*"

The best way to stand up for Creation is to be a good student first. There's a great deal of evidence against Evolution and this book only shows some of the large parts of that picture. You should study the reasons Creation scientists believe the Bible, but don't refuse to learn about Evolution. That will only make people think you are a fool. Be ready to talk about Creation, but be careful not to be a know-it-all either. If you stay kind and calm in your comments, others just might listen. God says, "*Come here and talk with me. Even though your sins are as red as scarlet, I will make them as white as snow.*" (Isaiah 1:18, paraphrased). God loves people and He wants them to know the truth. You should love people too, as you speak the truth to them.

How to remember this:

Evolution is a MYTH and you know that is true, but you should always speak kindly as you would have others speak to you.

M.Y.T.H.

- **M** = MEAN: being mean-spirited, rude or obnoxious is not a Godly way of talking to others about Evolution and Creation.
- **Y** = YIELD the chance to speak first to the other person. This is being "quick to listen, slow to speak and slow to get angry".
- **T** = TAKE IN - all you can learn about Evolution and Creation so you are a good student and an informed speaker instead of just emotional.
- **H** = HELPFUL, instead of arguing, you can help people consider the facts carefully and lower a big wall keeping them from listening to the Spirit of God. Remember, it is God who draws people to Himself, not our best arguments.

Notes

Chapter 1

[1] Goldschmidt, Richard B., "Evolution, As Viewed by One Geneticist." *American Scientist* vol.40 (January 1952): p. 84. Print.

Chapter 2

[2] Simpson, George Gaylord. "The Nonprevalence of Humanoids." *Science, vol.* 143 (February 21, 1964): 769. Print.

[3] Darwin, Charles. *The Autobiography of Charles Darwin*, 1887. As republished by the Norton Library: p. 94. Print. "A man who has no assured and ever present belief in the existence of a personal God...can have for his rule of life, as far as I can see, only to follow those impulses and instincts which are the strongest or which seem to him the best ones."

Chapter 3

[4] Matthews, L. Harrison (F.R.S.), Introduction to *On the Origin of Species.* London: J.M. Dent & Sons, Ltd., 1971. Print. "Belief in the theory of Evolution is thus exactly parallel to belief in special Creation: both are concepts that believers know to be true but neither, up to the present, has been capable of proof."

[5] Romans 1:20 (New King James Version): "For since the Creation of the world His invisible attributes are clearly seen, being understood by the things that are made, even His eternal power and Godhead, so that they are without excuse..."

Chapter 4

[6] Wolfs, F. 1996. "Introduction to the scientific method." *Physics Laboratory Experiments, Appendix E*, Department of Physics and Astronomy, University of Rochester. Web 1996. "No matter how elegant a theory is, its predictions must agree with experimental results if we are to believe that it is a valid description of nature. In physics, as in every experimental science, 'experiment is supreme' and experimental verification of hypothetical predictions is absolutely necessary."

Chapter 5

[7] Morris, Henry M. *The Genesis Record.* Grand Rapids, Michigan: Baker Book House, 1976. Print. p. 21 - 22. "The New Testament is, if anything, even more dependent on Genesis that the Old. There are at least 165 passages in Genesis that are either directly quoted or clearly referred to in the New Testament...Furthermore, in not one of these many instances where the Old or New Testament refers to Genesis is there the slightest evidence that the writers regarded the events or personages as mere myths or allegories. To the contrary, they viewed Genesis as absolutely historical, true, and authoritative."

[8] Morris, Henry M. *In the Beginning.* Green Forest, AR: Master Books, 1996. Print. "The Hebrew word for 'days' (yamin - the plural of yom, 'day'), which is used over 700 times in the Old Testament, never in any other place necessarily means anything but literal 'days.'"

[9] Morris, Henry M. *The Genesis Record.* Grand Rapids, Michigan: Baker Book House, 1991 edition. Print. p. 56. "The writer [of Genesis] not only defined the term day (Hebrew yom), but emphasized that it was terminated by a literal evening and morning and that it was like every other day in the normal sequence of days. In no way can the term be legitimately applied here to anything corresponding to a geologic period or any other such concept."

Chapter 7

[10]Morris, Henry M., Whitcomb, John. *The Genesis Flood.* Philipsburg, New Jersey: Presbyterian and Reformed Publishing, 1961. Print. p. 160.

[11] Gish, Duane T. *Evolution: The Fossils Still Say No.* El Cajon: Institute for Creation Research, 1995. Print. p. 187 "Thus bats appear in the fossil record fully-formed without a trace of ancestors or transitional forms, and they have remained essentially unchanged for the supposed 50 million years since they first appeared in the fossil record. This evidence is absolutely contradictory to Evolutionary theory but is precisely what is predicted on the basis of creation.

[12] Gish, Duane T. *Evolution: The Fossils Still Say No.* El Cajon: Institute for Creation Research, 1995. Print. P. 141 "It is only because of their desperate lack of transitional forms that Evolutionists have trumpeted so loudly about Archaeopteryx. Archaeopteryx appears abruptly in the fossil record, a powered flyer with wings of the basic pattern and proportions of the modern avian wing, and feathers identical to those of modern flying birds, an undoubted true bird without a single structure in a transitional state."

[13] Ibid, p. 139 "The duck-billed platypus is a strange mosaic, a definite mammal also possessing both reptilian and bird-like characteristics. As such it could not possibly be the ancestor or descendant of any other creature"

[14] Morris, Henry M., Morris, John. *The Modern Creation Trilogy, Vol 2.* Green Forest, AR: Master Books, 1996. Print. P.107. "Two hundred years of systematic fossil collecting, with billions of fossils now known, yet all the supposed Evolutionary links are still missing links! The laws of science demonstrate that Evolution is impossible, and the fossil record demonstrates that it never occurred even if it were possible."

Chapter 8

[15] Behe, Michael J. *Darwin's Black Box.* New York, Touchstone: 1996. Print. p. 39 "By irreducibly complex I mean a single system which is composed of several well-matched, interacting parts that contribute to the basic function, and where the removal of any one of the parts causes the system to effectively cease functioning."

[16] Sherwin, Frank. "Butterflies vs. Macroevolution". *Origins Issues.* Institute for Creation Research., 2010. Web. Feb 2005. "Detailed metamorphosis studies done on different insect species shows only that the ways organ systems are formed are quite varied. This is not what Evolution would predict.

[17] Milton, R., *Shattering the Myths of Darwinism*, Park St. Press: 1992. p. 220. Quoted from Sherwin, Frank. "Butterflies vs. Macroevolution". *Origins Issues*, Institute for Creation Research., 2010. Web. Feb 2005. ". . . no stage or aspect of this physical process can be accounted for or even guessed at with our current knowledge of chemistry, physics, genetics, or molecular biology, extensive though they are. It is completely beyond us. We know practically nothing about the plan or program governing the metamorphosis, or the organizing agency that executes this plan.

[18] Osorio, D, J. P. Bacon, and P. Whittington. 1997. The Evolution of arthropod nervous system. American Scientist vol. 95, p. 244. As quoted from Sherwin, Frank. "Butterflies vs. Macroevolution". *Origins Issues*, Institute for Creation Research., 2010. Web. Feb 2005. "As Darwin noted in the Origin of Species, the abrupt emergence of arthropods in the fossil record during the Cambrian presents a problem for Evolutionary biology. There are no obvious simpler or intermediate forms–either living or in the fossil record–that show convincingly how modern arthropods evolved from worm-like ancestors. Consequently there has been a wealth of speculation and contention about relationships between the arthropod lineages.

[19] Gish, Duane T. *The Amazing Story of Creation*. Green Forest, AR, New Leaf Press: 1996. Print. P. 53 "The Evolutionary scenario could never produce a butterfly by changing a caterpillar into a mass of jelly. But God, the master engineer of the universe, programmed a mass of jelly to develop into a delicately sculptured butterfly."

[20] Behe, Michael J. *Darwin's Black Box*. New York, Touchstone: 1996. Print. p. 187 "In the face of the enormous complexity that modern biochemistry has uncovered in the cell, the scientific community is paralyzed. No one at Harvard University, no one at the National Institutes of Health, no member of the National Academy of Sciences, no Nobel prize winter - no on at all can give a detailed account of how the cilium, or vision, or blood clotting, or any complex biochemical process might have developed in a Darwinian fashion."

[21] Ibid. p. 193. "Life on earth at its most fundamental level, in its most critical components, is the product of intelligent activity. The conclusion of intelligent design flows naturally from the data itself - not from sacred books or sectarian beliefs. Inferring that biochemical systems were designed by an intelligent agent is a humdrum process that requires no new principles of logic or science."

Chapter 9

[22] Morris, Henry M. "Probability and Order vs Evolution". *Impact* Institute for Creation Research, 2010. Web. July 1, 1979

[23] Cooper, Mary Ann - Associate Professor, Departments of Emergency Medicine and Bioengineering, University of Illinois at Chicago. lightningsafety.noaa.gov/medical.htm. National Weather Service. Web. 2010.

[24] Gish, Duane T. *The Amazing Story of Creation*. Green Forest, AR, New Leaf Press: 1996. Print.p. 35 "Sir Fred Hoyle said that the probability of Evolution is equal to the probability that a tornado, sweeping through a junkyard, would assemble a Boeing 747."

[25] H.P. Yockey. Journal of Theoretical Biology. Vol 76: pp. 337-343, 1977 as quoted in Gish, Duane T. *Creation Scientists Answer Their Critics*. El Cajon: Institute for Creation Research, 1993. Print. p. 265. "Yockey's conclusions are that the probability of a functional cytrochrome C evolving by random process is only 2 x 10-94, far, far below the impossibility threshold...a probability so low that it wouldn't happen once in the entire universe in 20 billion years, even if every star in the universe (over 100 billion) had a planet like the earth and one hundred amino acids could be assembled one trillion times a second on every planet."

[26] Hoyle, Sir Fred, and Chandra Wickramasinghe, *Evolution from Space*. New York: Simon & Schuster: 1984. Print. p. 148 "The likelihood of the spontaneous formation of life from inanimate matter is one to a number with 40,000 noughts after it...It is big enough to bury Darwin and the whole theory of evolution. There was no primeval soup, neither on this planet nor on any other, and if the beginnings of life were not random, they must therefore have been the product of purposeful intelligence."

[27] Morris, Henry M. "Probability and Order vs Evolution". *Impact* Institute for Creation Research, 2010. Web. July 1, 1979 "Whatever effect selection may possibly have had on random processes in later biological reproduction, it is clear beyond any rational argument that chance processes could never have produced even the simplest forms of life in the first place. Without a living God to create life, the laws of probability and complexity prove beyond doubt that life could never come into existence at all.

Chapter 10

[28] Asimov, Isaac. "In the Game of Energy and Thermodynamics You Can't Even Break Even," *Smithsonian Institute Journal* June 1970: pp. 4 - 11. Print. P. 6 "This law is considered the most powerful and most fundamental generalization about the universe that scientists have ever been able to make." P. 11 "In fact, all we have to do is nothing, and everything deteriorates, collapses, breaks down, wears out, all by itself - and that is what the Second Law is all about."

[29] Morris, Henry. *That their words may be used against them.* El Cajon,CA, Institute for Creation Research: December 1997. Print. p. 65 "The two laws of thermodynamics are the most universally applicable and impregnably confirmed of all laws of science."

[30] Morris, Henry. *Scientific Creationism.* El Cajon, CA, Master Books: March 1987, p. 43 - 45. Print.

[31] Simpson, George Gaylord, and W.S. Beck. *Life: An Introduction to Biology, 2nd ed.* New York: Harcourt, Brace & World, Inc.: 1965. Print. p.466. "But the simple expenditure of energy is not sufficient to develop and maintain order. A bull in a china shop performs work, but he neither creates nor maintains organization. The work needed is particular work; it must follow specifications; it requires information on how to proceed."

[32] Blum, Harold F. *Time's Arrow and Evolution.* Princeton University Press: 1968. p. 119 "No matter how carefully we examine the energetics of living systems we find no evidence of defeat of thermodynamic principles, but we do encounter a degree of complexity not witnessed in the nonliving world."

[33] Morris, Henry. *Scientific Creationism.* El Cajon, CA, Master Books: March 1987, p. 43 - 45. Print. P. 45 "Until Evolutionists can not only speculate, but demonstrate, that there does exist in nature some vast program to direct the growth toward higher complexity of the marvelous organic space-time unity known as the terrestrial biosphere (not to mention that of the cosmos), as well as some remarkable global power converter to energize the growth through converted solar energy, the whole Evolutionary idea is negated by the Second Law."

Chapter 11

[34] Muller, H.J. "Radiation Damage to the Genetic Material," *American Scientist,* vol. 38 January 1950: pp. 35. Print. "But mutations are found to be of a random nature, so far as their utility is concerned. Accordingly, the great majority of mutations, certainly well over 99 per cent, are harmful in some way, as is to be expected of the effects of accidental occurrences."

[35] Ayala, Francisco J. "Teleological Explanations in Evolutionary Biology," *Philosophy of Science,* vol. 37 March 1970: pp. 1-15. Print. "It is probably fair to estimate the frequency of a majority of mutations in higher organisms between one in ten thousand, and one in a million per gene per generation....mutation provides the raw materials of Evolution."

[36] Morris, Henry M. *Scientific Creationism,* Green Forest, Arkansas. Master Books, 1985. p.136. Print. "As a matter of fact, the creation model does not, in its basic form, require a short time scale.

It merely assumes a period of special creation sometime in the past, without necessarily stating when that was. On the other hand, the Evolution model does require a long time scale."

[37] Stansfield, William D. *The Science of Evolution*. New York: Macmillan, 1977, p. 84. Print. "It is obvious that radiometric techniques may not be the absolute dating methods that they are claimed to be. Age estimates on a given geological stratum by different radiometric methods are often quite different (sometimes by hundreds of millions of years). There is no absolutely reliable long-term radiological 'clock.'"

[38] Snelling, Andrew. "The Age of Australian Uranium". *Creation Ex Nihilo*, vol. 4, No. 2, 1981: pp.44 - 47. Print. "A logical extension of these data and conclusions is to suggest, as others already have, that [radioactive] ratios may have nothing to do with the age of a mineral."

[39] Morris, John. *The Young Earth*. Green Forest, AR: Master Books, 1994, p. 55. Print.

[40] Morris, Henry. *Scientific Creationism*. El Cajon, CA, Master Books: March 1987, p.139. Print. "Thus, at best, apparent ages determined by means of any physical process are educated guesses and may well be completely unrelated to the true ages."

Chapter 12

[41] Morris, John. *The Young Earth*. Green Forest, AR: Master Books, 1994, p.41 Print. Quoting Wald, George, *The Origin of Life, in Physics and Chemistry of Life*, 1955, p. 12 "...the old Earth concept is a necessary part of Evolution. Everyone agrees that Evolution is an unlikely process, involving millions and millions of favorable mutations, fortuitous environmental changes, etc. Only as one shrouds Evolution in the mists of time does it become respectable. If the Earth is billions of years old, there is enough time for unlikely events to occur, or so it is thought."

[42] Morris, Henry. *Scientific Creationism*. El Cajon, CA, Master Books: March 1987, p. 136. Print. "As a matter of fact, the creation model does not, in its basic form, require a short time scale. It merely assumes a period of special creation sometime in the past, without necessarily stating when that was. On the other hand, the Evolution model does require a long time scale."

"Although the creation model is not necessarily linked to a short time scale, as the Evolution model is to a long scale, it is true that it does fit more naturally in a short chronology."

Morris, Henry M., Whitcomb, John. *The Genesis Flood*. Philipsburg, New Jersey: Presbyterian and Reformed Publishing, 1961. p. 125. Print.

[44] Austin, Steve, *Impact* Institute for Creation Research, April 1, 1983. Print.

[45] Whitcomb, John. *The World that Perished* Baker Book House, 1988. p. 86. Print.

[46] Morris, Henry M., Whitcomb, John. *The Genesis Flood*. Philipsburg, New Jersey: Presbyterian and Reformed Publishing, 1961. p. 209. Print. "In light of such frequent flagrant contradictions [to uniformitarianism]...and also in addition to the innumerable evidences of catastrophe, rather than uniformity...the writers feel the data of geology do not provide valid evidence against the historicity of the universal deluge as recorded in the book of Genesis."

[47] Morris, John. *The Young Earth*. Green Forest, AR: Master Books, 1994, p. 83. Print. "Dividing the known amount of helium by the rate of accumulation shows that all of the helium in the atmosphere today would have accumulated in no more than two million years."

[48] Morris, Henry. *Scientific Creationism*. El Cajon, CA, Master Books: March 1985, p. 153. Print. "The significant thing to note, however, is that in every case the calculated apparent age of the ocean is vastly less than the supposed 5 billion year age of the Earth."

[49] Morris, Henry. *Scientific Creationism.* El Cajon, CA, Master Books: March 1985, p. 160. Print. "Finally, we mention again the fact that there are many processes which give young ages than processes which give old ages. Contrary to popular opinion, the actual facts of science do correlate better and more directly with a young age for the Earth than with the old Evolutionary belief that the world must be billions of years in age."

Chapter 13

[50] Ager, Derek V., *The New Catastrophism.* Cambride, U.K.: Cambridge University Press, 1993. p. xi. Print. "On that side, too, were obviously untenable views of bible-oriented fanatics, obsessed with myths such as Noah's Flood, and of classicists thinking of Nemesis. That is why I think necessary to include the following 'disclaimer': in view of the misuse that my words have been put to in the past, I wish to say that nothing in this book should be taken out of context and thought in any way to support the views of the 'creationists' (who I refuse to call 'scientific')."

[51] Morris, Henry M. *The Genesis Record.* Grand Rapids, MI: Baker Book House, 1976. pp. 21 - 22. Print. "Furthermore, in not one of these many instances where the Old or New Testament refers to Genesis is there the slightest evidence that the writers regarded the events or personages as mere myths or allegories. To the contrary, they viewed Genesis as absolutely historical, true, and authoritative."

[52] Whitcomb, John C. *The World that Perished.* Grand Rapids, MI: Baker Book House, 1988. p. 27. Print.

[53] Morris, Henry M., Whitcomb, John. *The Genesis Flood.* Philipsburg, New Jersey: Presbyterian and Reformed Publishing, 1961. p. 87 "it is by no means unreasonable to assume that all land animals in the world today have descended from those which were on the Ark."

[54] Ibid. p. 66. "Over 500 varieties of the sweet pea have been developed from a single type since the year 1700."

[55] Woodmorappe, John. *Noah's Ark: A Feasibility Study.* Santee, CA: Institute for Creation Research, 1996. p. 7. Print. "If, as the preponderance of the evidence shows, the created kind was equivalent to the family (at least in the case of mammals and birds), then there were only about 2,000 animals on the ark."

[56]. Ibid. p. 7. "In order to make this exercise more interesting, I have deliberately made the problem of animal housing on the Ark more difficult by adopting the genus as the taxonomic rank of the created kind. This necessitates, as shown below, nearly 16,000 animals on the Ark."

[57] Ibid. p. 13. "As can be seen from the census the vast majority of the animals on the Ark were small. Without allowing representation of large animals as juveniles, the median animal on the Ark would have been the size of a small rat (about 100 grams)...from the tabulation, it can be seen that only about 11% of the animals on the Ark were substantially larger than sheep."

[58] Ibid. p. 16 "Less than half the cumulative area of the Ark's three decks need to have been occupied by the animals and their enclosures." P. 20 "merely 3 - 6 thousand cubic meters of volume, which is 6 - 12% of the interior Ark volume, sufficed for the 371 day supply of food for the 16,000 animals."

[59] Ibid. entire study.

[60] Ibid. p xi. "It is shown that it was possible for 8 people to care for 16,000 animals, and without miraculous Divine intervention."

Question 1

[61] Burbidge, Geoffrey, "Why Only One Big Bang?" *Scientific American*. Feb 1992: p.120. Print. "Big Bang cosmology is probably as widely believed as has been any theory of the universe in the history of Western civilization. It rests, however, on many untested, and in some cases untestable, assumptions. Indeed, big bang cosmology has become a bandwagon of thought that reflects faith as much as objective truth."

[62] Darling, David, "On Creating Something from Nothing," *New Scientist*, Vol 151. Sept 14, 1996: p. 49. Print. "You cannot fudge this by appealing to quantum mechanics. Either there is nothing to begin with, in which case there is no quantum vacuum, no pre-geometric dust, no time in which anything can happen, no physical laws that can effect a change from nothingness into somethingness; or there is something, in which case that needs explaining."

[63] Margon, Bruce, "The Missing mass," *Mercury*. January/February 1975: p. 2. Print. "But observations indicate that only ten to twenty percent of the required mass is in the form of galaxies. The rest is unaccounted for and the understanding of this discrepancy is likely to alter some of our fundamental concepts of the universe."

[64] Oldershaw, Robert L., "What's Wrong with the New Physics?" *New Scientist*, vol. 128. December 22/29, 1990: p. 59. Print. "First, the big bang is treated as an unexplainable event without a cause. Secondly, the big bang could not explain convincingly how matter got organized into lumps (galaxies and clusters of galaxies). And thirdly, it did not predict that for the Universe to be held together in the way it is, more than 90 per cent of the Universe would have to be in the form of some strange, unknown dark form of matter...Theorists also invented the concepts of inflation and cold dark matter to augment the big bang paradigm and keep it viable, but they too have come into increasing conflict with observations."

[65] Hoyle, Sir Fred, "The Big Bang under Attack," *Science Digest*, vol. 92. May 1984: p. 84. Print. "I have little hesitation in saying that a sickly pall now hangs over the big-bang theory. When a pattern of facts becomes set against a theory, experience shows that the theory rarely recovers."

Question 2

[66] Morris, Henry M., Morris, John. *The Modern Creation Trilogy, Vol 2*. Green Forest, AR: Master Books, 1996: p. 105. Print. "Can anyone seriously believe that the first bird really hatched out of a reptile egg? Actually, there would have to be at least two such hopeful monsters - one male and one female - occurring simultaneously in the same population, in order to assure survival of the new type. It would seem that one could as easily believe in a fairy godmother with a magic wand!"

[67] Ibid. p. 105 "Considerable resistance to this punctuated equilibrium concept is evident...but it has probably become the dominant view. If Evolutionists were really willing to take the scientific evidence seriously, they would have to conclude that neither gradual upward Evolution nor saltational upward Evolution has ever occurred, and that neither ever could occur."

Question 3

[68] Strobel, Lee. *The Case for a Creator*. Grand Rapids, MI: Zondervan, 2004: p. 132. Print. "One expert said there are more than thirty separate physical or cosmological parameters that require precise calibration in order to produce a life-sustaining universe." Reference to Stephen C. Meyer, "Evidence for Design in Physics and biology" in Michael J. Behe, William A. Dembski, and Stephen C. Meyer, editors, *Science and Evidence for Design in the Universe*. San Francisco: Ignatius, 2000:, p. 60. Print.

[69] Collins, Robin as quoted in Strobel, Lee. *The Case for a Creator*. Grand Rapids, MI: Zondervan, 2004: p. 133 - 134. Print. "Let's say you were way out in space and were going to throw a dart at

random toward the Earth. It would be like successfully hitting a bull's eye that's one trillionth of a trillionth of an inch in diameter. That's less than the size of one solitary atom."

[70] Ibid. p. 130. "When scientists talk about the fine-tuning of the universe they're generally referring to the extraordinary balancing of the fundamental laws and parameters of physics and the initial conditions of the universe. Our minds can't comprehend the precision of some of them. The result is a universe that has just the right conditions to sustain life."

[71] Gish, Duane T. *The Amazing Story of Creation*. Green Forest, AR, New Leaf Press: 1996: p. 24 "There are many, many other things on the Earth that have to be exactly the way they are for life to exist. All those things just couldn't happen by accident. That is one reason why many scientists, who don't believe the Bible, still believe in creation."

[72] Strobel, Lee. *The Case for a Creator*. Grand Rapids, MI: Zondervan, 2004: p. 151. Print. "...the everyday functioning of the universe is, in itself, a kind of ongoing miracle. The "coincidences" that allow the fundamental properties of matter to yield a habitable environment are so improbably, so far-fetched, so elegantly orchestrated, that they require a divine explanation."

Question 4

[73] DeYoung, Donald. *Astronomy and the Bible*. Grand Rapids, MI: Baker Book House, 1988: p. 46 - 47. Print. "At present there is no physical evidence for such a comet reservoir, nor indeed any way to verify its existence. Regardless, many astronomers have a secular faith that the Oort cloud exists and that it ensures a long time scale for the solar system. Perhaps there is an alternative: the presence of comets may be an evidence that the solar system is not nearly as old as is assumed by many people."

Question 5

[74]Thomas, Brian. "Martian lake still won't lead to life." *Daily science update*, Institute for Creation Research, 2010. Web. July 10, 2009. "But water is only one of dozens of necessary preconditions for sustaining life, including "a hydrologic cycle coexisting with land,... a strong magnetic field;" a system for balancing carbon dioxide, oxygen, ethane, and methane; specified radioactive decay rates; continents to mix marine nutrients; just the right size and mass for the planet; and a satellite very much like the Moon that can stabilize the Earth's tilt and support vital marine algal oxygen production by causing tides."

[75] DeYoung, Donald. Whitcomb, John. *The Moon: Its Creation, Form and Significance*. Winona Lake, IN: BMH books, 1978: pp. 43 - 44. Print. "Hammond has given some details on the constraints for lunar capture. (Hammond, "Exploring the Solar System (III): Whence the Moon?" 912) He explains that in order for the Moon to be captured intact, its speed could not exceed 40 meters/second. If its speed exceeded 2,500 meters/second, it would be diverted into a new heliocentric orbit, not captured at all. In between these two extremes it would have broken into particles and distributed itself into Saturn-type rings...There is simply no known means by which the Moon's velocity could be largely dissipated on a single pass."

[76] Ibid. p. 46,48. "A crucial problem with this view is the precise balance needed during the proposed buildup of the proto-Earth and proto-Moon...However, this proposal begins to resemble the already-existing belt of asteroid particles beyond the orbit of Mars, which certainly shows no tendency to condense into a single solid. One begins to wonder how such naturalistic theories can endure in the face of such enormous problems."

[77] Morris, Henry M., Morris, John. *The Modern Creation Trilogy, Vol 2*. Green Forest, AR: Master Books, 1996: p. 231"The physical and chemical composition of the Moon is very different from that

of the Earth, however, and it is difficult to see how the Moon could have come from the Earth, even as the result of such a hypothetical giant collision. Though the impact explanation is currently favored – for lack of anything better – it is surely not a very good one."

Question 7

[78] Anonymous, "Hoyle on Evolution," *Nature*, vol. 294. November 12, 1981: p. 105. Print. "The essence of his argument last week was that the information content of the higher forms of life is represented by the number 10 40,000 – representing the specificity with which some 2000 genes, each of which might be chosen from 10 20 nucleotide sequences of the appropriate length, might be defined. Evolutionary processes would, Hoyle said, require several Hubble times to yield such a result. The chance that higher life forms might have emerged in this way is comparable with the chance that 'a tornado sweeping through a junk-yard might assemble a Boeing 747 from the materials therein.'"

[79] Morris, Henry M., Morris, John. *The Modern Creation Trilogy, Vol 2*. Green Forest, AR: Master Books, 1996: p. 193. Print. ""The famous experiment of Miller, who was able to produce certain amino acids in his laboratory under conditions supposedly simulating conditions on the primitive Earth, has frequently been cited as of importance almost equal to the work of Charles Darwin. His experiment was entirely artificial, of course, and come nowhere near to producing life in the laboratory, as many of the credulous have been led to believe. Miller's conclusion has been completely negated now by the rapidly growing knowledge that the primeval Earth did not have the required reducing atmosphere at all."

[80] Ibid. p. 192 "Now, however, like so many other Evolutionary fables, the primeval "reducing" atmosphere is being dissipated by the hard facts of science. For more than a decade now, various scientists have been developing a wide range of evidences that the Earth's atmosphere was rich in oxygen right from the beginning!"

[81] McCombs, Charles, PhD. "Evolution Hopes You Don't Know Chemistry: The Problem with Chirality." *Impact*, Institute for Creation Research, 2010. Web. May 1, 2004. " In our body, every single amino acid of every protein is found with the same left-handed chirality. Although Miller and Urey formed amino acids in their experiments, all the amino acids that formed lacked chirality. It is a universally accepted fact of chemistry that chirality cannot be created in chemical molecules by a random process."

[82] Crick, Francis. *Life Itself: Its Origin and Nature.* New York: Simon & Schuster, 1981: p. 88. Print. "An honest man, armed with all the knowledge available to us now, could only state that in some sense, the origin of life appears at the moment to be almost a miracle, so many are the conditions which would have had to have been satisfied to get it going."

Question 8

[83] Henry Morris, John Whitcomb. The Genesis Flood. The Presbyterian and Reformed Publishing Company, 1961. P. 153 (in footnote, figure 6). "According to uniformist concepts, numerous changes of environment, with great regional subsidences and uplifts, must have been involved, but this would appear quite impossible. The strata simply could not have remained so nearly uniform and horizontal over such great areas and great periods of time, while undergoing such repeated epeirogenic movements."

[84] Morris, John. *The Young Earth.* Green Forest, AR: Master Books, 1994: p. 97. Print. "Each rock stratum immediately overlies another, with no soil between, even though vast lengths of time are presumed to have passed between the two."

[85] Ibid. p. 94. "(Raindrops and animal tracks) had to be formed in soft sediment, are very fragile, and if present on any surface, will not last very long."

[86] Austin, Steven A. "Mt. St. Helens and Catastrophism" *Impact*, Institute for Creation Research. 2010. Web. July 1, 1986. "The little "Grand Canyon of the Toutle River" is a one-fortieth scale model of the real Grand Canyon. The small creeks which flow through the headwaters of the Toutle River today might seem, by present appearances, to have carved these canyons very slowly over a long time period, except for the fact that the erosion was observed to have occurred rapidly! Geologists should learn that, since the long-time scale they have been trained to assign to landform development would lead to obvious error on Mount St. Helens, it also may be useless or misleading elsewhere."

[87] Mehlert, Albert. "Diluviology and Uniformitarian Geology, a Review." *Creation Research Society Quarterly*, Vol. 23, No. 3. December 1986: p. 106. Print. "A week's study of the Grand Canyon should be a good cure for Evolutionary geologists as it is a perfect example of flood geology...the whole area was obviously laid down quickly, then uplifted and then the whole sediment were split open like a rotten watermelon."

Question 9

[88] Sherwin, Frank. "Running Counter to Evolution." *Origins Issues*, Institute for Creation Research. 2010. Web. Sept 1, 2004. "These exchangers in the tongue contain a network of small veins surrounding larger and warmer arterial vessels coming from the body core. The veins contain cooler, venous blood going toward the heart by way of the jugular vein. Dozens of these heat exchangers allow the shift of heat from warm blood coming from the body's interior to the adjacent venous blood returning from the extremities. As a result of this exchange, only a small amount of the animal's heat is lost to the cold waters (see Science 278:1138-1139)."

[89] Gish, Duane T. *Evolution: The Fossils Still Say No*. El Cajon: Institute for Creation Research, 1995: p. 207. Print. ""It is clear that the evidence weighs heavily on the creationist side concerning the origin of marine mammals. It requires an enormous faith in miracles to believe that some hairy, four-legged mammal crawled into the water and gradually, over eons of time, gave rise to whales, dolphins, sea cows, seals, sea lions, walruses, and other marine mammals via thousands and thousands of random genetic errors."

[90] Thomas, Brian. "Museum's 'Science' exhibit leaves more questions than answer." *Daily Science Updates*, Institute for Creation Research. 2010. Web. Jan 11, 2010. "But what is not revealed is that Ambulocetus, illustrated as a possible transitional "link" to the modern whale, is missing its key body part, "since the pelvic girdle is not preserved." Nor did the display accurately show that another link in the sequence, Basilosaurus, was ten times larger than Ambulocetus, or that many Evolutionists believe Basilosaurus does not even belong in modern whale Evolutionary ancestry. Remarkably, Pakicetus was also on display. Although it was depicted as a distant whale ancestor when only skull fragments of it were known, more recent data shows it had four legs that were fully fitted for land-life!"

[91] Gish, Duane T. *Evolution: The Fossils Still Say No*. El Cajon: Institute for Creation Research, 1995: p. 205. Print. "Robert Carroll, in his voluminous tome, Vertebrate Paleontology and Evolution, made the incredible statement that "Despite the extreme difference in habitus, it is logical from the standpoint of phylogenetic classification (how we place animals in order) to include the mesonychids (wolf-like Mesonyx) among the Cetacea (whales). Presto! These wolf-like animals are now whales! Who says Evolutionists do not have transitional forms? Anybody who can call a wolf a whale should have no trouble finding "transitional forms."

[92] Ibid. p 208. "The gaps in the fossil record between the major kinds are systematic and usually large. The transitional forms, especially those that must be visualized as leading up to highly specialized mammals, such as bats, are not found in the fossil record."

[93] Morris, Henry M., Morris, John. *The Modern Creation Trilogy, Vol 2*. Green Forest, AR: Master Books, 1996: p. 51. Print. "Though some gaps certainly are to be expected, because of the accidental nature both of fossilization and of fossil discoveries, such gaps should at least be randomly distributed between the present kinds of animals and the transitional forms...if the fossil gaps were not regular and systematic but only statistical - all of this would be hailed as striking, indisputable proof of Evolution....Since, however, this type of evidences is altogether lacking in the fossil record, auxiliary conditions have to be imposed on the Evolutionist predictions."

Question 10

[94] Hoyle, Sir Fred. *The Intelligent Universe*. New York: Holt, Rinehart & Winston, 1983: p. 23. Print. "In short there is not a shred of objective evidence to support the hypothesis that life began in an organic soup here on the Earth."

[95] Morris, Henry M., Morris, John. *The Modern Creation Trilogy, Vol 2*. Green Forest, AR: Master Books, 1996: p. 58. Print. "It does seem odd that these and other "billion-year" old organisms, which tend to multiply quite rapidly, would remain so unaffected by Evolutionary pressures for so long, if Evolution is true. Presumably, however, at least one did change, because the rocks of the Cambrian "period" (or system) team with multitudes of marine invertebrate animals of great complexity and fantastic variety - worms, brachiopods, starfish, jellyfish, trilobites, and a host of others...Yet, out of the billons of fossils of Precambrian protozoa and billons of Cambrian metazoan fossils, nowhere has there ever been found a transitional form from a Precambrian one-celled animal to a Cambrian many-celled invertebrate."

[96] Gish, Duane T. *The Amazing Story of Creation*. Green Forest, AR, New Leaf Press: 1996: p. 44, 45. Print. "If Evolutionists are correct, these Cambrian animals appeared on Earth about 2.5 billion years after the first microscopic form of life evolved...the truth is, not one single fossil intermediate or transitional form, has ever been found."

[97] Gish, Duane T. *Creation Scientists Answer Their Critics*. El Cajon: Institute for Creation Research, 1993: p. 119. Print. "Surely, if paleontologists are able to find numerous fossils of microscopic, single celled, soft-bodied bacteria and algae, then they could easily find fossils of all stages intermediate between these microscopic organisms and the complex invertebrates of the Cambrian. Fossils of thousands of these intermediate stages should grace museum displays. None have been found."

[98] Gish, Duane T. *Evolution: The Fossils Still Say No*. El Cajon: Institute for Creation Research, 1995: p. 69. Print. "What greater evidence for creation could the rocks give than this abrupt appearance of a great variety of complex creatures without a trace of ancestors? Thus we see, right from the beginning, on the basis of an Evolutionary scenario, the evidence is directly contradictory to predictions based on Evolution but is remarkable in accord with predictions based on creation."

Question 11

[99] Maisey, J. (1996) *Discovering Fossil Fishes*. Henry Holt & Co., p. 121 as cited in Sherwin, Frank, "Fish that talk." *Back to Genesis*. Institute for Creation Research. 2010. Web, June 1, 1999. "In cladistic analysis, humans are osteichthyans, because tetrapods—craniates with arms and legs—evolved from a group of lobe-finned fishes that in turn evolved from the true bony fishes, the osteichthyans. Humans have been happy to classify themselves as primates . . . but they intuitively

shy away from regarding themselves as bony fishes. Why omit this intermediate stage from our pedigree? . . . We should come to terms with the idea that we belong to a highly specialized group of bony fishes."

[100] Sherwin, Frank, "Fish that talk." *Back to Genesis.* Institute for Creation Research. 2010. Web, June 1, 1999. "Thought extinct for 80-100 million years and used as an index fossil, the coelacanth (Latimeria) was found, alive and well, off South Africa. Recently, this "progenitor of the human race" has also been found off Sulawesi, Indonesia. Evolutionists had maintained that the coelacanth, with its four lobed fins, resembled primitive tetrapod limbs and was the forerunner to the first land animals. The living specimens were investigated. What researchers found proved disappointing for their cause. The fins are ordinary fish fins with cartilage, not bone. Furthermore, the fins are structured in such a way that they couldn't possibly become legs. In 1986 and again in 1996 (Copeia, v. 3, p. 607) the coelacanth was filmed using its fins only for swimming, just as creationists predicted. Soft tissue anatomy, such as the brain, heart, and intestine, were not what Evolution theory predicted would be found. This fish speaks of no change, stasis, in the face of Darwinism that demands transition."

[101] Morris, Henry M., Morris, John. *The Modern Creation Trilogy, Vol 2.* Green Forest, AR: Master Books, 1996: p. 114. Print. "Other such publicized living fossils include the Metasequoia dawn redwood tree (previously thought to be extinct since the Miocene epoch, 20 million years ago), the tuatara, or beakhead reptile (supposedly extinct since the Cretaceous), the segmented mollusk Neopilina (extinct it was thought, since the Devonina, 300 million years ago), and the brachiopod shellfish Lingula (presumed extinct for about 400 million years, since the Ordovician)."

[102] Morris, John. "Did lungfish evolve into amphibians." *Dr. John's Q and A.* Institute for Creaiton Research. 2010. Web. July 1, 1996. "A majority of today's evolutionists hold to the idea that a similar type of fish (order, *Rhipidistia),* led to amphibians. Again, this fossil fish had structures in its fins, and a loose comparison could be made with the femur and humerous (arm and leg bones in land animals), but nothing to compare to hands and feet. Furthermore, as is also the case in the coelacanth, the hard parts of the fins are loosely embedded in muscle, not at all attached to the vertebra as required to support the weight of the body."

[103] *Science News.* Vol. 102 (9/21/72): p. 152. "Throughout the hundreds of millions of years, the coelacanths have kept the same form and structure. Here is one of the great mysteries of Evolution."

Question 12

[104] Kitts, David. "Paleontology and Evolutionary Theory," *Evolution*, vol. 28. Sept. 1974: p.466. Print. "Evolution, at least in the sense that Darwin spoke of it, cannot be detected within the lifetime of a single observer!"

[105] Morris, John. "What about the peppered moth?" *Dr. John's Q and A.* Institute for Creation Research. 2010. Web. April 1, 1999. "Remember that both varieties were present at the start, with the mix of genes producing lights favored over the mix of genes producing darks. As the environment changed, the dark variety had greater opportunity to pass on their genetic mix, and percentages changed. All the while, the two types were interfertile. No new genes were produced, and certainly no new species resulted. This is natural selection in action, but not Evolution. Adaptation happens, but the changes are limited."

[106] Morris, Henry. *Many Infallible Proofs.* Green Forest, AR: Master Books, 1974: p. 254. Print. "The peppered-moth...is a good case in point. The basic kind was still unchanged - the moth was still a moth. Natural variation with selection is thus a conservative process, enabling a given kind of organism to be preserved even though the nature of its habitat has been altered."

[107] Morris, Henry. "Evolutionists and the Moth Myth." *Back to Genesis*. Institute for Creation Research. 2010. Web. Aug. 1, 2003. "One of the main questioners has been Ted Sargent, emeritus professor of Biology at the University of Massachusetts, who insists that the famous photographs of moths on tree trunks published by Kettlewell were all fakes...We cannot discuss all these criticisms here, but the conclusion was, as Hooper says: ". . . at its core lay flawed science, dubious methodology, and wishful thinking" (Hooper, p. xx). Some went so far as to accuse Ford and Kettlewell of actual fraud, but most thought it was just poor science. Cambridge lepidopterist, Michael Majerus, in his book, Melanism: Evolution in Action "left no doubt that the classic story was wrong in almost every detail" (Hooper, p. 283)."

[108] Morris, Henry. *Scientific Creationism*. El Cajon, CA, Master Books: March 1985: p. 52. Print. "Since the creator had a purpose for each kind of organism created, He would institute a system which would not only insure its genetic integrity but would also enable it to survive in nature. Otherwise even very slight changes in its habitat, food supply, etc., might cause its extinction."

Question 13

[109] Woodmorappe, John. *Noah's Ark: A Feasibility Study*. Santee, CA: Institute for Creation Research, 1996: p.4. Print. "After all, it is they (Evolutionists) who suggested that dinosaurs were too heavy to walk on land without their massive bodies being supported by water. This notion had been widely prevalent in both popular and scientific depictions of dinosaurs (in fact, Bakker 1986, p.116, called it an "eighty-year orthodoxy"). Bakker has shown that the dinosaurs' anatomy is completely incompatible with even a swampy habitat."

[110] Morris, Henry M., Morris, John. *The Modern Creation Trilogy, Vol 2*. Green Forest, AR: Master Books, 1996: p. 121. Print. "Tales about dragons have come down from many nations all over the world, and it is simply impossible that these various peoples were all inventing the same imaginary animals. There must have been a basis of fact in their tales..."

[111] Gish, Duane T. *The Amazing Story of Creation*. Green Forest, AR, New Leaf Press: 1996: p. 67.

[112] Ibid. p. 98 "All of this happens extremely fast in the bombardier beetle's combustion tubes, heating the liquid and gases up to 212 degrees Fahrenheit, and generating a lot of pressure...A pop can actually be heard as the gases shoot out!"

Question 14

[113] Thomas, Brian. "Fossil fibers befuddle Evolution." *Daily Science Update*. Institute for Creation Research. 2010. Web. March 31, 2009. "No dinosaur has been found with any undisputed transitional feature. Even the newly-discovered "dino-fuzz" is not feather-like (lacking rachis, barbs, barbules, or evidence of keratin rather than cartilage) and is perhaps more accurately interpreted as simply decorative fibers."

[114] R. Monastersky, "All mixed up over birds and dinosaurs," *Science News* 157:38. (January 15, 2000). Print. "Red-faced and downhearted, paleontologists are growing convinced that they have been snookered by a bit of fossil fakery from China. The "feathered dinosaur" specimen that they recently unveiled to much fanfare apparently combines the tail of a dinosaur with the body of a bird, they say."

[115] Gish, Duane T. *Evolution: The Fossils Still Say No*. El Cajon: Institute for Creation Research, 1995: p. 141. Print. "*Archaeopteryx* was an undoubted true bird without a single structure in a transitional state."

Question 15

[116] Gish, Duane T. *Evolution: The Fossils Still Say No*. El Cajon: Institute for Creation Research, 1995: p. 197. Print. "Hyrocotherium is not any more like a horse than it is similar to a tapir or a rhinoceros, and thus just as justifiably it could have been chosen as the ancestral rhinoceros or tapir... the "horse" on which the entire family tree of the horse rests was not a horse at all."

[117] Ibid. p. 193. "Thus, if Evolutionists permit the fossil evidence to be their guide, they should suppose that in South America a one-toed ungulate gave rise to a three-toed ungulate...this is precisely the opposite of the supposed sequence of events that occurred with North American horses."

[118] Simpson, George Gaylord. *Life of the Past*. New Haven: Yale University Press, 1953: p. 125. "The uniform continuous transformation of Hyrocotherium (Eohippus) into Equus, so dear to the hearts of generations of textbook writers, never happened in nature."

Question 16

[119] Ayala. Francisco, "The Mechanisms of Evolution," *Scientific American*. vol. 239, No. 3, 1978: pp. 56-69. Print.

[120] Parker, Gary. “Creation, Mutation, and Variation.” *Impact*. Institute for Creation Research. 2010. Web. Nov 1, 1980. “If we use Ayalas figures, there would be no problem at all. He cites 6.7 % as the average proportion of human genes that show heterozygous allelic variation, e.g., straight vs. curly hair, Ss. On the basis of "only" 6.7 % heterozygosity, Ayala calculates that the average human couple could have 10^{2017} children before they would have to have one child identical to another! That number, a one followed by 2017 zeroes, is greater than the number of sand grains by the sea, the number of stars in the sky, or the atoms in the known universe (a "mere" 10^{80})!”

[121] Morris Henry M., Whitcomb, John. The Genesis Flood. Baker Book House, 1989. P. 87. "It is by no means unreasonable to assume that all land animals in the world today have descended from those which were on the ark."

Question 17

[122] Gish, Duane T. *Creation Scientists Answer Their Critics*. El Cajon: Institute for Creation Research, 1993: p. 134. Print. "The track record of Evolutionists, in their search for man's fossil ancestry, has been a dreary record of failures and fraud."

[123] Lubenow, Marvin. *Bones of Contention*. Grand Rapids, MI: Baker Book House, 1992: p. 32. Print. "There is an abundance of hominid fossils, the bulk of them are either too modern to help me (the Evolutionist) or they do not fit well into Evolution's scheme...The reality is that by 1969 - 1976, approximately 4,000 hominid fossil individuals had already been unearthed...(since that time) a conservative estimate of the total number of hominid fossil individuals discovered to date exceeds 6,000."

[124] Lubenow, Marvin. *Bones of Contention*. Grand Rapids, MI: Baker Book House, 1992: p. 137. Print.

[125] Gish, Duane T. *Evolution: The Fossils Still Say No*. El Cajon: Institute for Creation Research, 1995: p. 281. Print. "To have revealed this fact at that time would have rendered it difficult, if not impossible, for his Java man to have been accepted as a "missing link." His failure to reveal this find to the scientific world...was deplorable since this constituted concealment of important evidence."

[126] Ibid. P. 296. "Sinanthropus (Peking Man) consisted of the skulls of either large macaques (large monkeys) or large baboons killed and eaten by workers at an ancient quarry."

[127] Lubenow, Marvin. *Bones of Contention.* Grand Rapids, MI: Baker Book House, 1992: p. 138. Print. "My own conclusion is that *Homo erectus* and Neanderthal are actually the same."

[128] Gish, Duane T. *Evolution: The Fossils Still Say No.* El Cajon: Institute for Creation Research, 1995: p. 329. Print. "A modern apes jaw and a human skull had been doctored to resemble an ape-man, and the forgery had succeeded in fooling most of the world's greatest experts."

[129] Ibid. p. 328. "In 1927, after further collecting and studies had been carried out, it was decided that Hesperopithicus was neither a manlike ape nor an apelike man, but was an extinct peccary, or pig. This is a case in which a scientist made a man out of a pig, and the pig made a monkey out of the scientist!"

[130] Ibid. p. 331."There is thus no evidence, either in the present world or in the world of the past, that man has arisen from some "lower" creature. He stands alone as a separate and distinct created type, or basic morphological design, endowed with qualities that sets him far above all other living creatures."

Question 18

[131] Gish, Duane T. "Man...Apes...Australopithicines...Each Uniquely different." *Impact.* Institute for Creation Research. 2010. Web. Nov 1, 1975. "The dental evidence cited is the fact that, although these creatures, believed to have weighed 60-70 pounds, had cheek teeth as large as those found in 400-pound gorillas, thus possessing massive jaws, their front teeth (incisors and canine teeth) were relatively small in comparison with their cheek teeth when compared to the relative size of incisors and canines to cheek teeth in modern apes."

[132] Gish, Duane T. *Evolution: The Fossils Still Say No.* El Cajon: Institute for Creation Research, 1995: p. 243. Print. "According to the analysis of Johanson and White...Lucy and her fellow creatures walked upright in the human manner, although they were essentially ape-like from the neck up."

[133] Ibid. P. 239. "Lord Zuckerman's conclusion is that *Australopithecus* was an ape, in no way related to the origin of man...From his results, Oxnard concluded that *Australopithecus* did not walk upright in human manner."

[134] Ibid. P 256. "It may be that *Australopithecus* Afarensis and other Australopithicines were actually no more adapted to a bipedal mode of locomotion than are chimpanzees and Gorillas, which do occasionally walk bipedally."

[135] Lubenow, Marvin. *Bones of Contention.* Grand Rapids, MI: Baker Book House, 1992: pp. 154 - 155. Print.

[136] Ibid. p.174. "(the Laetoli prints are) indistinguishable from those of habitually barefoot Homo Sapiens."

[137] Ibid pp. 178 – 179 "Fossils that are indistinguishable from modern humans can be traced all the way back to 4.5 million years ago, according to the Evolutionists time scale. That suggests that true humans were on the scene before the Australopithicines appear in the fossil record. In other words, even when we accept the Evolutionary dates for the fossils, the results do not support human Evolution."

Question 19

[138] Gish, Duane T. *Evolution: The Fossils Still Say No.* El Cajon: Institute for Creation Research, 1995: p. 305. Print. "If he were given a shave, a haircut and a bath and dressed in a business suit,

and were to walk down one of our city streets, he would be given no more attention than any other individual."

[139] Ibid. P. 305. "It is known that these people suffered severely from rickets, caused by a deficiency of vitamin D. This condition results in softening of bone and consequent malformation."

[140] Morris, Henry M., Morris, John. *The Modern Creation Trilogy, Vol 2*. Green Forest, AR: Master Books, 1996: p. 90. "Neanderthal man, of course, is now acknowledged by all evolutionary anthropologists to be true man, Homo Sapiens."

[141] Ibid. p. 91 "Neanderthal's somewhat stooped posture, heavy skull, and other physical peculiarities may have been the result of disease or dietary deficiencies - possibly even old age - but they certainly were within the range of modern human attributes."

Question 20

[142] Morris, Henry. *The Biblical Basis for Modern Science*. Grand Rapids, MI: Baker Book House, 1984: p. 36. Print. "This leads inevitable to a choice between two alternatives (either the universe is) 1) an infinite chain of non-primary causes (there is no starting or ending point) or 2) (there is) an uncaused primary cause of all causes."

[143] Lewis, C.S. (Editor: Walter Hooper). *The Business of Heaven*. Harvest Books, Harcourt Brace and Co., 1984. "If anything emerges clearly from modern physics, it is that nature is not everlasting. The universe had a beginning, and will have an end.

[144] Morris, Henry. *The Biblical Basis for Modern Science*. Grand Rapids, MI: Baker Book House, 1984: p. 37. Print. "The first cause must be an infinite, eternal, omnipotent, omnipresent, omniscient, living, conscious, volitional, moral, spiritual, esthetic, loving being! Furthermore, since the universe is not a multiverse, the God who created it could only have been one God, not two gods or many gods."

Question 21

[145] Lewis, C.S. Mere Christianity. Macmillan Publishing Company. 1980: p. 43. Print.

[146] Morris, Henry. *The Biblical Basis for Modern Science*. Grand Rapids, MI: Baker Book House, 1984: p. 39. Print. *The God who is Real*, Grand Rapids, MI: Baker Book House, 1988: p. 53. Print. "God is not capricious...there must, therefore, be good and sufficient reason why He created the universe and man to live in the universe. It is reasonable, therefore, to assume that God would reveal his purposes by some means of clear communication to his creatures."

[147] McDowell, Josh. *A Ready Defense*. San Bernardino, CA: Here's Life Publishers, 1990: p. 28. Print. "No other document of antiquity even begins to approach such numbers and attestation. In comparison, the Iliad by Homer is second with only 643 manuscripts that still survive."

[148] Morris, Henry. Clark, Martin. *The Bible Has The Answer*. El Cajon, CA: Master Books, 1987: p. 2. Print. "There is no other book, ancient or modern, like this. The vague, and usually erroneous, prophecies of people like Jeanne Dixon, Nostradamus, Edgar Cayce, and others like them are not in the same category at all, and neither are other religious books such as the Koran, the Confucian analects, and similar religious writings. Only the Bible manifests this remarkable prophetic evidence, and it does so on such a tremendous scale as to render completely absurd any explanation other than divine revelation."

Question 22

[149] Romans 3:23 (NLT) "For all have sinned; all fall short of God's glorious standard."

[150] Isaiah 53:6 (NLT) "All of us have strayed away like sheep. We have left God's paths to follow our own. Yet the LORD laid on him the guilt and sins of us all."

[151] Romans 6:23 (NLT) "For the wages of sin is death, but the free gift of God is eternal life through Christ Jesus our Lord."

[152] Revelation 21:8 (NLT) But cowards who turn away from me, and unbelievers, and the corrupt, and murderers, and the immoral, and those who practice witchcraft, and idol worshipers, and all liars ~ their doom is in the lake that burns with fire and sulfur. This is the second death."

[153] Ephesians 2:8 (NLT) "God saved you by his special favor when you believed. And you can't take credit for this; it is a gift from God."

[154] Romans 6:9 (NLT) "We are sure of this because Christ rose from the dead, and he will never die again. Death no longer has any power over him. **10** He died once to defeat sin, and now he lives for the glory of God. **11** So you should consider yourselves dead to sin and able to live for the glory of God through Christ Jesus."

[155] Isaiah 1:18 (NLT) "Come now, let us argue this out," says the LORD. "No matter how deep the stain of your sins, I can remove it. I can make you as clean as freshly fallen snow. Even if you are stained as red as crimson, I can make you as white as wool."

[156] 1 John 4:10 (NLT) "This is real love. It is not that we loved God, but that he loved us and sent his Son as a sacrifice to take away our sins."

[157] Romans 3: 21 - 25 (NLT) "But now God has shown us a different way of being right in his sight ~ not by obeying the law but by the way promised in the Scriptures long ago. **22** We are made right in God's sight when we trust in Jesus Christ to take away our sins. And we all can be saved in this same way, no matter who we are or what we have done. **23** For all have sinned; all fall short of God's glorious standard. **24** Yet now God in his gracious kindness declares us not guilty. He has done this through Christ Jesus, who has freed us by taking away our sins. **25** For God sent Jesus to take the punishment for our sins and to satisfy God's anger against us. We are made right with God when we believe that Jesus shed his blood, sacrificing his life for us. God was being entirely fair and just when he did not punish those who sinned in former times."

[158] Romans 10: 9 - 10 (NLT) "For if you confess with your mouth that Jesus is Lord and believe in your heart that God raised him from the dead, you will be saved. **10** For it is by believing in your heart that you are made right with God, and it is by confessing with your mouth that you are saved."

[159] 1 John 1: 8 - 9 (NLT) "If we say we have no sin, we are only fooling ourselves and refusing to accept the truth. **9** But if we confess our sins to him, he is faithful and just to forgive us and to cleanse us from every wrong."

[160] Romans 10:13 (NLT) "For "Anyone who calls on the name of the Lord will be saved."

[161] Hebrews 10:24 - 25 (NLT) "Think of ways to encourage one another to outbursts of love and good deeds. 25 And let us not neglect our meeting together, as some people do, but encourage and warn each other, especially now that the day of his coming back again is drawing near."